# SAFE DESIGN AND CONSTRUCTION
## OF MACHINERY

# Safe Design and Construction of Machinery
## Regulation, Practice and Performance

ELIZABETH BLUFF

*The Australian National University, Canberra, Australia*

CRC Press
Taylor & Francis Group
Boca Raton London New York

CRC Press is an imprint of the
Taylor & Francis Group, an **informa** business

CRC Press
Taylor & Francis Group
6000 Broken Sound Parkway NW, Suite 300
Boca Raton, FL 33487-2742

© 2015 by Elizabeth Bluff
CRC Press is an imprint of Taylor & Francis Group, an Informa business

No claim to original U.S. Government works

Printed on acid-free paper
Version Date: 20160616

ISBN-13: 978-1-4724-5077-7 (hbk)
ISBN-13: 978-1-1388-9291-0 (pbk)
ISBN-13: 978-1-3156-0741-2 (ebk)

**Visit the Taylor & Francis Web site at**
**http://www.taylorandfrancis.com**

**and the CRC Press Web site at**
**http://www.crcpress.com**

# About the Author

Elizabeth Bluff has 30 years' experience in occupational health and safety (OHS) and risk management. She is a Director of the National Research Centre for OHS Regulation in the Regulatory Institutions Network, at The Australian National University where she also holds an appointment as a Research Fellow. She has a Bachelor of Science (Hons) from the University of Adelaide, a Masters of Applied Science (OHS) from the University of Ballarat and a PhD in OHS regulation from Griffith University, Queensland.

# Contents

# List of Figures and Tables

**Figures**

**Tables**

# List of Abbreviations

| | |
|---|---|
| ABS | Australian Bureau of Statistics |
| ANSI | American National Standards Institute. |
| ASCC | Australian Safety and Compensation Council |
| ASSE | American Society of Safety Engineers |
| AWCBC | Association of Workers' Compensation Boards of Canada |
| BSI | British Standards Institution |
| CE | Certification Europe |
| CEN | European Committee for Standardization |
| CENELEC | European Committee for Electrotechnical Standardization |
| CSA | Canadian Standards Association |
| DIISRTE | Department of Industry, Innovation, Science, Research and Tertiary Education (Australia) |
| DTI | Department of Trade and Industry (UK) |
| EC | European Community |
| FEM | Federation of European Materials Handling |
| HSE | Health and Safety Executive (UK) |
| HWSA | Heads of Workplace Safety Authorities (Australia) |
| ILO | International Labour Organization |
| IRC | Industrial Relations Commission |
| ISO | International Organization for Standardization |
| KAN | Kommission Arbeitsschutz und Normung (Germany) |
| NIOSH | National Institute for Occupational Safety and Health (US) |
| NOHSC | National Occupational Health and Safety Commission (Australia) |
| OSHA | Occupational Safety and Health Administration (US) |
| OHS | Occupational health and safety |
| PPE | Personal protective equipment |
| SAA | Standards Association of Australia |
| SWEA | Swedish Working Environment Authority |
| UK | United Kingdom |
| US | United States |
| VTHC | Victorian Trades Hall Council |
| WRMC | Workplace Relations Ministers' Council (Australia) |

# Preface

This book deals with the inter-related themes of risk, regulation, business practice and performance, based on an in-depth study of safety in the design and construction of machinery. Why machinery – because globally it takes a heavy toll in work-related deaths and injuries. Why design and construction – because removing hazards and integrating control measures at the source is one of the most cost-effective ways to manage risks. This is well recognized in the growing number of public policy and regulatory initiatives addressing health and safety problems 'upstream', including the Australian and European regulatory regimes for machinery safety in focus in this research.

The book is, however, much more than an account of business performance and responses to regulation in a particular context. It takes a fresh look at capacity and motivation as central elements shaping business conduct, and their highly contextualized nature. It offers insights into the impact of state regulation alongside the influence of non-state actors in firms' supply chains and networks.

This means that the book will appeal to an international audience from diverse backgrounds – those interested in human factors and safety engineering, work and product safety, risk management, regulation and socio-legal studies, sociology of work, standard setting and enforcement, and professional or vocational education. And, across these multiple fields, readers may come to the book as researchers, specialists or practitioners, regulators and policy makers, educators or students.

The book itself is multidisciplinary. I hope that by integrating literature and theory from different disciplines with empirical findings about safety in machinery design and construction, the book will help build bridges between specialist, regulatory and practitioner bodies of knowledge and communities of practice. It is my belief that only by applying a multidisciplinary perspective to understanding how and why health and safety problems arise, can we hope to develop and implement effective solutions.

While conducting the research and writing this book I have been working at the National Research Centre for OHS Regulation (NRCOHSR), which is part of the Regulatory Institutions Network (RegNet) at the Australian National University. The manuscript undoubtedly benefitted from discussions with my NRCOHSR and RegNet colleagues, and members of our wider networks. I would like to thank especially four people who read and provided very welcome and constructive feedback on the manuscript at different stages of its development. They are: Professor Richard Johnstone at the Australian Centre for Health Law Research, Queensland University of Technology; Professor Bridget Hutter at the Centre for Analysis of Risk and Regulation, London School of Economics; John Braithwaite,

Distinguished Professor and founder of RegNet; and my work health and safety colleague and very good friend Dr Clare Gallagher.

I am also indebted to the 66 manufacturing firms and the 32 staff of the occupational health and safety (OHS) regulators that participated in the research, and gave their time generously to contribute their understandings and experiences of safety in machinery design and construction, regulation and compliance. Finally I would like to thank my partner Des for his continuing encouragement and invaluable insights into the realities of industrial working life, which have also helped to shape my understanding of health and safety, and its implementation in practice.

Elizabeth Bluff
*Canberra*

# Chapter 1

# Introduction

What shapes business performance for social goals such as safety? Why are some firms' products inherently safe while others endanger safety? How do state imposed legal obligations and enforcement influence business conduct, and how does their influence compare with that of non-state actors in global markets and supply chains? What role do knowledge and motivational factors play in shaping firms' actions and performance for safety, and how are they constituted? Are specialist bodies of knowledge, such as those for human factors and safety engineering, applied in practice? What are the implications of all of this for safety policy and practice?

These are some of the significant social issues discussed in this book. They are topics that span the interests of researchers, regulators and policy makers, specialists, practitioners, educators and students across multiple fields in safe design, human factors and safety engineering, work and product safety, risk management, regulation and socio-legal studies, sociology of work, standard setting, and professional and vocational education, among others. The book offers readers from these diverse perspectives fresh insights into business responses to public policy, regulatory and professional imperatives, through an in-depth study of risk management in machinery design and construction. The research blends different literatures and theoretical approaches with empirical investigations to enrich understanding of how, to be effective in regulating and managing risks, we need to pay greater attention to the real nature of work and corporate life, and appreciate the complex contextual influences that shape business conduct.

The rationale for examining safety in machinery design and construction stems from the heavy toll that machinery takes globally in work deaths and injuries. Statistical data are not directly comparable between countries but as a broad indication, each year in the European Union machinery is a contributing factor in more than 300,000 injuries, which is 11 per cent of all injuries involving more than three days off work (European Commission, 2008), while machinery is an even more prominent cause of work injuries in China where 30 per cent of injuries treated in hospital emergency departments are machinery-related (Fitzharris, et al., 2011). Annually there are at least 65,000 injuries involving days away from work in the United States (Harris and Current, 2012), 15,000 injuries involving time off work in Canada (AWCBC, 2012), and around 3,500 hospitalizations from machinery-related injuries in Australia (Safe Work Australia, 2009; 2011; 2013a).

From hand-held power tools to complex production systems, machinery may pose genuine and serious risks to health and safety. Most well recognized are mechanical hazards as the following, not uncommon, examples illustrate:

A machine operator was fatally crushed in a machine. He had entered the service area of a production line to clear an obstruction, triggering an automatic safety device, which stopped the machine. The machine was turned on again by an operator who sat at a console, in a position from which he could not see the operator in the service area.

A farm worker suffered fatal injuries when his jacket caught on the auger of a drilling rig, pulling him into the machine. There was no caging around the drill, interlock or dead-man control on the operating panel. (Examples from NOHSC, 2000, pp. xiii, 86).

As well as the inadequately guarded danger zones and poorly positioned controls that these examples highlight, machinery may be hazardous through weak structures that collapse or break apart, hazardous chemical emissions and leaks, noise and vibration, the ergonomic problems of awkward postures or repetitive movements in machinery operation, and complex human–technology interfaces that give risk to mental strain, human error and hazardous incidents (Al-Tuwaijri, et al., 2008; Backstrom and Döös, 1997; 2000; Brauer 1994; 2006; Gardner, et al., 1999). There is also compelling evidence that a high proportion of machinery-related deaths and injuries are attributable to its poor design and construction in the first instance (Driscoll, et al., 2005; 2008; NIOSH, 2013; Safe Work Australia, 2009, p. 15).

The importance of inherently safe design has been recognized in a series of public policy and professional initiatives in the United States, Europe and Australia, based on the premise that one of the most effective ways to prevent work-related deaths and injuries is to design out hazards and integrate risk control measures at the source (ASSE, 2011; European Commission, 2008, pp. 209–10; Kletz, 1998a; 1998b; Manuele, 1999a; 2008; NOHSC, 2002; Safe Work Australia, 2012a,b; Swuste, 1997; Schulte, et al., 2008). There is also a substantial specialist body of knowledge, originating in the disciplines of human factors and safety engineering, to support the structured analysis and resolution of safety problems from early in the life cycle of machinery (for example Brauer, 1994; 2006; Corlett and Clark; 1995; Green and Jordan, 1999; Karwowski, 2005; Karwowski and Marras, 1999; Morris, Wilson and Koukoulaki, 2004; Stanton and Young, 1999; Stanton, et al., 2005).

On the legal side, the pre-eminent regulatory regime requiring the safe design and construction of machinery is the law of member states in the European Union giving effect to the *Machinery Directive* (European Commission, 1998a; 2006). Australian occupational health and safety (OHS) law also has a well-developed framework of legal obligations for machinery designers and manufacturers (Bluff, 2004; Johnstone, 1997, pp. 260–3; 2004a, pp. 275–80). In other countries the OHS legal obligations of employers may be the impetus for machinery producers to conform to safety standards, as with the American National Standards Institute

(ANSI) standards for safeguarding machinery (Harris and Current, 2014; OSHA, 2014).

The empirical research presented in this book was conducted with Australian firms that manufactured and supplied a wide variety of machinery into international markets, as well as locally. By virtue of the transnational application of the policy, professional and regulatory imperatives outlined above, and the international scope of the literature underpinning them, this research has relevance for an international readership grappling with issues of safety in the design and construction of products, and business responses to policy and regulatory interventions more generally.

In illuminating the mechanisms underlying manufacturers' responses for machinery safety the research also makes wider conceptual and theoretical contributions. It provides insights into knowledge and motivational factors as principal elements shaping firm performance for social and regulatory goals, and advances understanding of how these elements are constituted in the everyday operations of firms and their interactions with external actors.

## Overview of the Research

The research presented in this book focused on the design and construction of machinery, as distinct from supply or end use, because in the earlier life cycle stages there is the opportunity to produce machinery that is inherently safer.[1] This can be achieved if those making decisions about design and construction choose structures, materials and components which eliminate hazards, and risk control measures that are integral to the design, compatible with machine functionality, and hence less likely to be removed or disarmed (Kletz 1998a; 1998b; Polet, Vanderhaegen and Amalberti, 2003; Reunanen, 1993, p. 108; Swuste, et al., 1997; Seim and Broberg, 2010).

Centre stage in the research are the European regulatory regime for machinery safety based on the *Machinery Directive* (European Commission, 1998a; 2006), and the obligations of designers and manufacturers in Australian OHS law (Bluff, 2004; Johnstone, 1997, pp. 260–3; 2004a, pp. 275–80). These are leading examples of state regulatory requirements for the safe design and construction of machinery, and the regimes most applicable to the study firms. Although by no means harmonized, the European and Australian regimes contain some common elements. Among these are obligations for the management of risks and provision

---

1 In later life cycle stages preventive measures are limited to retrofitting risk control measures, avoiding risks arising through poor installation or post-production modifications, and regular servicing and maintenance to help ensure the integrity and efficacy of existing control measures. While these are important preventive measures they do not make machinery inherently safer.

of safety information, and both regimes are underpinned by detailed technical standards[2] for particular types or aspects of machinery.

Taking a wider, de-centred view of regulation the research also examined the non-state actors in local, national and transnational domains that influenced business conduct (see also Black, 2001a; Hutter and Jones, 2007; Parker and Nielsen, 2009). For machinery manufacturers, the state and non-state actors differ according to each firm's operations and markets. They might include state regulators or policy bodies in a firm's home or export countries, national and international standards bodies, business contacts in supply chains or networks (suppliers of component parts, customers or distributors of end products), providers of education and training, professional bodies, industry and trade associations, unions and insurance companies, among others.

The research set a substantive goal of both the Australian and European regulatory regimes for machinery safety as the overarching benchmark of firm performance and compliance. This was the goal of preventing death, injury and illness (the regulatory goal of prevention). For prevention purposes it was critical that manufacturers comprehensively recognized hazards, eliminated those hazards or incorporated risk control measures to minimize the risks, and provided safety information that was accessible to and comprehensively informed end users about machinery safety matters. Keeping the regulatory goal of prevention clearly in focus, the research examined manufacturers' actions and standards of performance for the substantive safety outcomes of hazard recognition, risk control and provision of safety information, and the factors and processes shaping their responses.

The research design and methodology are set out in full in the Appendix. In brief, the sample for the empirical study with machinery manufacturers was drawn from firms in two Australian states (Victoria and South Australia), and included a cross-section of small, medium and large businesses,[3] in capital city and regional locations. Collectively the 66 firms in the sample produced more than 30 different types of machinery or equipment including various types of cranes, hoists and lifting equipment, agricultural and horticultural machinery, boilers and compressors, industrial cleaning systems, and an array of machinery for processing, handling or packaging food, beverages, wood, minerals, vehicles, and other products or waste materials. The study firms supplied their machinery in international markets in Europe, Asia, North America or the Middle East, as well as 15 different industry sectors around Australia. In each study firm, the informants were key individuals responsible for making and implementing decisions about machinery design and construction as directors, owners, or managers overseeing production, engineering and other technical or specialist functions.

---

2   Technical standards are published documents that establish detailed engineering or technical specifications or procedures; for example, European harmonized standards, Australian Standards and international (ISO) standards (see also Chapter 2).

3   Small = < 20 employees; medium = 20–99 employees; large = 100 or more employees.

A second empirical study with OHS regulators investigated their inspection and enforcement policy and practice for machinery design and construction. Data collection for the two empirical studies involved in-depth, face-to-face interviews in manufacturing firms and with the regulators, supplemented by review of documentation and, for manufacturers, observation of machinery to identify potential sources of harm and risk control measures incorporated or absent. The two empirical studies were underpinned by a legal review and analysis of the principal legal obligations (Australian and European), applying to the safe design and construction of machinery.

The research provided evidence of the mixed performance of manufacturers for hazard recognition, risk control and safety information. More importantly, the research contributed to understanding why some firms performed well for these safety outcomes while others failed to do so. It distinguished knowledge about machinery safety matters and motivational factors (motivations, values and attitudes) as the principal elements shaping firm action and, in turn, performance for substantive safety outcomes. The research also demonstrated the highly contextualized nature of knowledge about machinery safety matters and of motivational factors, as they were constituted in the operations of firms and through interactions with external actors. These external actors might help build capacity or spread misinformation, and they might motivate or constrain preventive action by manufacturers. Key decision makers in firms also shaped firm behaviour through the influence of their personal histories, values and attitudes. State regulation (Australian and European) contributed to the knowledge and motivation to address machinery safety in some firms but, even when state regulation had some influence it had to compete with other constituents of knowledge and motivations. As a consequence, firm behaviour was idiosyncratic and performance for substantive outcomes was often insufficient for firms to comply with the regulatory goal of prevention.

## Research Contributions

The research builds on the growing body of scholarship demonstrating the influence of the social and economic contexts of firms' operations on their compliance with state regulation, and performance for safety specifically. Examples of such scholarship are studies of business responses to safety-related legal obligations and enforcement (Fairman and Yapp, 2005a; 2005b; Genn, 1993; Gray and Scholz, 1993; Haines, 1997; Hutter, 2001; 2011; Kagan and Scholz, 1984; Mendeloff and Gray, 2005), and studies of business responses to social and economic regulation more generally (Braithwaite V, 2009; Braithwaite V, et al., 1994; Gunningham, Kagan and Thornton, 2003; Gunningham, Thornton and Kagan, 2005; May and Wood, 2003; Parker, 2002; and see generally Parker and Nielsen, 2011, and contributors therein).

The rich data generated for machinery manufacturers enabled a nuanced account of the principal elements shaping their performance for substantive safety outcomes. These elements are motivations, values and attitudes (motivational factors); knowledge about machinery safety; state regulation (Australian and European legal instruments and enforcement systems); and non-state institutions and actors in the form of technical standards bodies, parties in firms' supply chains and networks, and health and safety professionals. In turn this enabled the development of explanation and theory about the nature of and interplay between these elements and substantive safety outcomes, through inductive reasoning grounded in the empirical data, and deductive reasoning drawing on the literature to interpret the data and interrogate emerging explanation (Marshall and Rossman, 2006, pp. 161–2; Morse and Richards 2002, pp. 169–70; Neuman, 1997, pp. 46–8; Richards, 2005, pp. 128–34; Silverman, 2001, pp. 237–40).

At one level the empirical findings and theorizing from this research converge with and reinforce Parker and Nielsen's (2011, pp. 5, 9–26) conclusion that to explain business behaviour we must understand the influence of and interplay between the goals or priorities that motivate that behaviour, organizational capacities and characteristics that shape decision making and implementation, state regulation and enforcement, and non-state influences. At a deeper level the research makes conceptual and theoretical contributions to understanding and explaining motivational factors and knowledge, their constitution in the everyday operations of firms and interactions with external actors, and how these factors and processes shape firm behaviour and whether or not they comply with state regulation. In essence the research uncovers the web of influences that create plural responses among manufacturers and differentiate their performance for substantive safety outcomes.

The principal motivational factors for machinery manufacturers could be characterized, in a general way, as legal, economic or normative, but not social (see also Ayres and Braithwaite, 1992, pp. 23–5; Kagan, Gunningham and Thornton, 2011; May, 2004; Parker and Nielsen, 2011, pp. 10–12). It was, however, more useful to describe them precisely so as to reveal and make clear their origins and how they influenced firm behaviour. By exploring the mix of influential motivations, values and attitudes it also became clear that, within a particular firm, these factors might be mutually reinforcing or conflicting. Apparent drivers for firms to take action on machinery safety could be cancelled out by barriers to taking such action, as when an espoused moral obligation to protect human health and safety was counteracted by over-riding business concerns about the functionality and marketability of machinery. Moreover, when the relationships between motivational factors and substantive safety outcomes were examined it was evident that only some factors were actually linked with good performance for these outcomes; those that cast machinery safety as in some way integral to the success of the business. Others, for example manufacturers' reputational concerns, were simply espoused motivations, sometimes termed psychological compliance

(Parker and Nielsen, 2009, p. 57), which did not necessarily drive constructive preventive action.

With regard to knowledge, the research established that key individuals (key decision makers) in manufacturing firms constructed knowledge about machinery safety matters from multiple bases but differences in constituents of learning did not, in themselves, explain differences in knowledge about machinery safety. A social constructivist perspective of learning (Billett, 2001; Palincsar, 1998) was useful in theorizing learning about machinery safety in firms, as the research suggested that both social and individual processes were involved in the construction of knowledge about machinery safety. Key individuals constructed safety knowledge principally through participation in everyday activities and interactions with others, within and outside their businesses, and interpreted what they experienced through the lens of their personal domains of knowledge due to their different personal histories, capacities and agency (Billett 1996; 2001; 2003; 2008a; Scribner and Beach, 1993).

State regulation and enforcement were part of the mix of motivational factors and constituents of knowledge in manufacturing firms, although state regulatory demands were generally not well understood and were rather lost among other inputs with which they competed for authority. Intriguingly, specialist human factors and safety engineering sources, which offer information, methods and tools to support the integration of safety in design, were little used in study firms, and only in firms that employed or engaged human factors or other safety professionals. In contrast the influence of the wider external environment of non-state actors was considerable. This influence was represented in firms, and key individuals in firms, privileging parties in supply chains and their industry contacts as drivers of action and sources of information. Yet this research suggests caution in regulators relying on or harnessing such non-state actors to motivate or build the capacity of regulatees, as some safety and socio-legal scholars have proposed (see for example Gunningham and Sinclair, 2002, pp. 17–18; Hopkins and Hogan, 1998; Lamm and Walters, 2004, pp.103–5; Walters, 2001, pp. 52, 375–6; 2002, pp. 45–6). There was little evidence that market and industry influences were linked with manufacturers performing well for substantive safety outcomes and, in some respects, the findings signal the need for regulators to contemplate strategies to reshape or, in worst cases, disarm the counterproductive influence of some market and industry actors.

In essence then, knowledge and motivational factors shaped the quality and rigour of manufacturers' actions for machinery safety and, in turn, their performance for substantive safety outcomes. By their nature, the constituents of knowledge and motivational factors were highly contextualized, complex and unpredictable, as they arose from manufacturers' operations and the mix of practices and interactions in which firms, and key decision makers in firms, participated within and outside their businesses. In some firms, motivational factors, knowledge and actions constituted commitment, capacity and arrangements as the pre-conditions

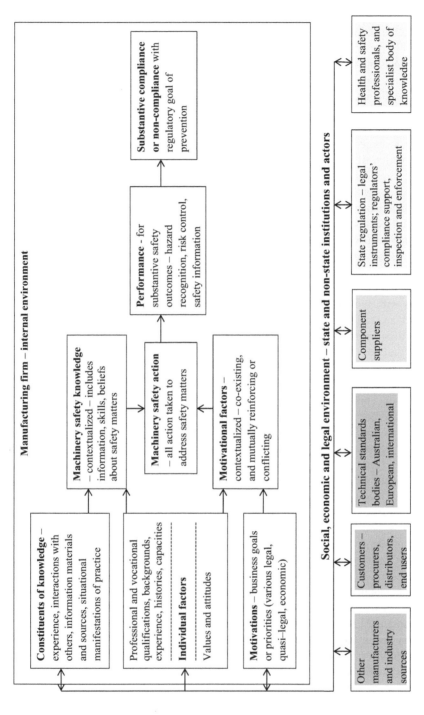

**Manufacturing firm – internal environment**

**Constituents of knowledge –** experience, interactions with others, information materials and sources, situational manifestations of practice

**Individual factors**

Professional and vocational qualifications, backgrounds, experience, histories, capacities

Values and attitudes

**Motivations** – business goals or priorities (various legal, quasi–legal, economic)

**Machinery safety knowledge** – contextualized – includes information, skills, beliefs about safety matters

**Machinery safety action** – all action taken to address safety matters

**Motivational factors** – contextualized – co-existing, and mutually reinforcing or conflicting

**Performance** - for substantive safety outcomes – hazard recognition, risk control, safety information

**Substantive compliance or non-compliance** with regulatory goal of prevention

**Social, economic and legal environment – state and non-state institutions and actors**

Other manufacturers and industry sources

Customers – procurers, distributors, end users

Technical standards bodies – Australian, European, international

Component suppliers

State regulation – legal instruments; regulators' compliance support, inspection and enforcement

Health and safety professionals, and specialist body of knowledee

**Figure 1.1    The plural constituents of manufacturers' responses for machinery safety**

for achieving sound safety performance and complying with the regulatory goal of prevention (on pre-conditions for self-regulation see Johnstone and Jones, 2006; Parker, 2002, p. ix–x, 43–61). In other firms the nature of motivational factors and knowledge impeded or constrained such preventive action.

Two examples illustrate how contextualized knowledge and motivational factors uniquely shaped manufacturers' responses. They show how manufacturers pursuing alternative business goals from different knowledge bases achieved very different standards of performance for substantive safety outcomes.

The first example is a manufacturer of surface finishing machines for use with acrylic, timber, marble, metal and other types of surfaces (Manufacturer 10 in the study). The principal base from which the key individuals in this firm constructed knowledge about machinery safety was their personal experience as end users of surface finishing machines. They had experienced the musculoskeletal strain, vibration, dust and other hazards with this type of machinery. Their knowledge informed the design of a new type of machine. For this firm producing a safe and ergonomically sound machine was a business opportunity, and hence machinery safety was integral to the success of the firm. As the business expanded, the firm's goal of supplying its machine to Europe was the motivation to comply with various European directives relating to machinery safety, which the director had learned about by contacting several business support and government agencies in Australia. In this firm, both knowledge and motivations supported the achievement of substantive safety outcomes. The firm produced an inherently safe machine, applied technology in original ways to address the key hazards, and provided substantial, good quality safety information in a booklet, video and labels on the machine.

The second manufacturer, which produced food processing systems, (Manufacturer 54 in the study) provides a contrasting example. The managing director drew upon his experience as an engineer in the chemical processing industry. He consulted the firm's customers, but not end users of the machinery. He had engaged a consultant to advise on certain aspects of machinery safety but had not sought information about, and had no knowledge of, the firm's legal obligations for machinery safety. The managing director's technical knowledge supported comprehensive hazard recognition and the application of advanced technology to control some risks but his key motivation, the marketability of the machinery, impeded rigorous attention to all safety risks. The firm did not incorporate risk control measures if the managing director perceived that they would reduce the functionality and hence the marketability of the machinery.

An important implication of the empirical findings and theorizing from this research is that as manufacturers made decisions within the framework of their contextualized knowledge and motivations, their decision making was often characterized by bounded rationality (Gigerenzer and Selten, 2001; Simon, 1955). This was evident, for example, in manufacturers' choice to adopt industry standard designs or control measures that were the product of copying other firms' machinery, and were compatible with their own functionality and marketability goals, but were not sufficient to ensure compliance with the regulatory goal of prevention. To meet the latter standard, firms needed to make well rounded decisions based on authoritative sources of information about hazards and ways of eliminating or minimizing risks, and they needed to implement preventive measures proportionate to risks.

## Structure of the Book

The book has nine chapters, which are presented in a sequence that reflects the unfolding account and explanation of the plural responses of machinery manufacturers. Following this Introduction, Chapter 2 looks at what the state expected of machinery manufacturers, analysing the principal legal obligations for the safe design and construction of machinery applicable to the study firms, distinguishing the core elements of these obligations and the elements that the different regulatory regimes have in common. Chapter 3 then evaluates manufacturers' performance for substantive safety outcomes relating to three core elements of their legal obligations – hazard recognition, risk control and provision of safety information. This evaluation indicates the level of compliance by firms with the regulatory goal of prevention, and sets the foundations for exploring, in the remainder of the book, the factors and processes that shaped firm performance.

Chapter 4 is the first of two chapters dealing with the influence of state regulation. The chapter examines awareness of the relevant legal obligations in manufacturing firms as, in principle, such obligations may contribute both to knowledge and motivations in firms. It presents empirical evidence of the generally low awareness of legal obligations, and appraises alternative explanations for this low level of awareness. Chapter 5 examines interactions between OHS regulators and machinery manufacturers. It documents the types of contacts that captured firms' attention, prompting them to take action on machinery safety matters. This chapter also unpicks the key reasons why, although armed with broad functions and powers, the OHS regulators were less than optimally effective in the arena of machinery design and construction.

The next part of the book moves beyond state regulation to examine other conceptual themes of interest in understanding and explaining manufacturer performance. Chapter 6 looks at how those involved in design and construction activities learned about safety matters. The chapter explains the importance of the specialist body of professional knowledge originating in the human factors and safety engineering disciplines, and examines the extent to which manufacturers engaged with this body of knowledge. The chapter then clarifies the more central role of firms' everyday practices and interactions with parties in their supply chains and networks in shaping safety knowledge in firms. Chapter 7 considers the extent to which firms had institutionalized arrangements to assess and manage risks in machinery design and construction, and the nature of risk assessment practice. Chapters 6 and 7 each reveal factors and processes linked with better, or poorer, firm performance for substantive safety outcomes. Chapter 8 completes this analysis, elucidating the plural motivations arising from legal and economic imperatives which, together with the values and attitudes of key individuals, sustained or constrained firms' action on machinery safety matters.

Chapter 9 consolidates the empirical, conceptual and theoretical contributions of the research. It advances theory to explain the relationship between knowledge and motivational factors in manufacturing firms, their action and performance for substantive safety outcomes, and compliance (or non-compliance), with the regulatory goal of prevention. The chapter also sets out some policy implications of the research, and proposes some strategic directions for state regulation and for capacity building.

# Chapter 2
# The Legal Obligations for Designing and Constructing Safe Machinery

The manufacturers studied faced a series of legal obligations for the safe design and construction of the machinery they produced. Put another way, there were important legal reasons why it would have been prudent for them to anticipate and take steps to minimize the risk of harm arising from their machinery.

Within their home country of Australia, the principal legal obligations were established in occupational health and safety (OHS) law as the firms produced a wide variety of machinery for use at work. The substantive goal of Australian OHS law was the prevention of death, injury and illness arising from work-related hazards and risks (Johnstone, 1997, pp. 100–2; 2004a, pp. 99–101). Provisions for the safe design and construction of machinery were contained in nine general OHS statutes, and the regulations and approved codes of practice made under these statutes, as each Commonwealth, state and territory government in Australia's federal system was empowered to make OHS laws (Johnstone, 2004a, pp. 87–90; Johnstone, Bluff and Clayton, 2012, pp. 92–3). As the machinery was not intended for and not likely to be used for personal, domestic or household purposes the separate law for consumer product safety did not apply (Johnstone 1997, p. 267; 2004a, p. 291).

For the firms that supplied machinery into international markets there were additional legal obligations in some of the countries to which they supplied. In particular those intending to put their machinery on the market in the European Union were required to comply with the law of member states giving effect to the *Machinery Directive*, and separate directives for specific types of machinery and for particular hazards (European Commission, 1998a; 1998b; 1998c; 2000a, pp. 12–13; 2006).[1]

There were 19 study firms that exported machinery to European countries including Britain, France, Germany, Greece, Spain and the Nordic countries. As a large producer and market for machinery, Europe's regulatory regime for machinery safety and the harmonized standards underpinning it have also been influential in international markets such as China and other countries in Asia (European Commission, 2014; European Standards Organizations, 2009; IMS Research, 2009; and see Lacore, 2002). The European regime based on the *Machinery Directive* has two key goals: to facilitate the free movement of

---

1   The *Machinery Directive* was revised and reissued in 2006 for application in 2009 (Bamberg and Boy, 2008; Fraser, 2010).

machinery and safety components within Europe; and to prevent machinery that endangers health or safety from being placed on the market and put into service in Europe (European Commission, 1998a, art 2; 2000a, p. 7; 2006, art 4). It embodies the principle that the social cost of accidents caused by machinery can be reduced by inherently safe design and construction of machinery (European Commission, 1998a, recital 4; 2006, recital 2). The European regime applies both to machinery supplied for commercial use and that supplied for domestic use.

This chapter reviews and analyses the principal legal obligations, as they applied at the time of data collection for the research.[2] The first part of the chapter examines the relevant provisions of Australian OHS law. The chapter then considers the European regulatory regime for machinery safety. This comparative analysis is the basis for distinguishing the core elements of the state imposed legal obligations for the safe design and construction of machinery.

These analyses reveal that, notwithstanding differences in the detail of specific legal obligations, at the level of broad principles Australian OHS law and the European regulatory regime for machinery safety had (and continue to have) some core elements in common. The chapter finishes by explaining that some of these core elements were applied in this research, as a conceptual framework for examining manufacturers' performance and arrangements for addressing machinery safety matters.

## The Australian Regulatory Regime

*Overview of the Australian Legal Obligations*

Prior to the 1980s in Australia, statutory obligations for machinery design and construction applied only to specific types of machinery such as boilers and pressure vessels, lifts and cranes, or to machinery supplied to particular industries (Brooks, 1993, pp. 630–4, 687 and 736; Gunningham, 1984, pp. 195–7, 375–91; Worksafe Australia, 1996, pp. 24–36). From the 1980s, the 'new style' OHS statutes incorporated duties for persons who designed and/or constructed 'plant' which applied to producers of a wide range of machinery, equipment, appliances, implements and tools. These statutes drew extensively on the 1972 British Robens Committee's report, Safety and Health at Work which, in turn, was influenced by the *Federal German Law on Technical Equipment 1968*, and the Swedish *Workers' Protection Act 1949*, both of which required manufacturers of machinery to address the safety of their products (Johnstone, Bluff and Clayton, 2012, pp. 70–3; Robens, 1972, p. 112). For its part the Robens Committee proposed that

---

2   As well as the revision of the *Machinery Directive*, some of the Australian OHS statutes, regulations and approved codes of practice have been amended or remade subsequently.

machinery and equipment must, so far as practicable, be designed and constructed so as to be intrinsically safe in use (Robens, 1972, pp. 111–14).

Victoria was the first Australian state to extend a duty of care to designers and manufacturers of machinery and other types of plant[3] in its 1981 statute. By the early 1990s duties of manufacturers, and generally also designers, were included in all of the principal OHS statutes and have remained a feature in the current statutes (Johnstone, 1997, pp. 261–3; Johnstone, Bluff and Clayton, 2012, pp. 331–5, 352–8). At the time of data collection for this research the OHS statutes in seven jurisdictions (NSW, NT, Qld, SA, Tas, Vic, WA) incorporated duties for both designers and manufacturers of plant, and the OHS statutes for the other two jurisdictions (ACT, Cth) incorporated duties for manufacturers.[4] The duties were similar in scope across the nine jurisdictions although they differed in detail.

From the mid-1990s, the statutory general duties were reinforced by regulations and/or generic codes of practice based on the *National Standard for Plant* (NOHSC, 1994; WorkSafe Australia, 1996).[5] The regulations established mandatory requirements and the generic codes of practice could be used in legal proceedings as evidence of an acceptable standard of care, but failure to observe a code did not in itself render a person liable to criminal or civil proceedings (Johnstone, 1997, pp. 294–6). There were also some technical standards which were mandatory as they were prescribed in regulations, or evidentiary as they were

---

3    Australian OHS law uses the term 'plant' which is defined to include machinery, equipment, appliances, implements and tools.

4    The OHS statutes were: Australian Capital Territory – *Occupational Health and Safety Act 1989* (OHSA 1989 (ACT)); Commonwealth – *Occupational Health and Safety (Commonwealth Employment) Act 1991* (OHSA (CE) 1991 (Cth)); New South Wales – *Occupational Health and Safety Act 2000* (OHSA 2000 (NSW)); Northern Territory – *Work Health Act 1986* (*WHA 1986* (NT)); Queensland – *Workplace Health and Safety Act 1995* (WHSA 1995 (Qld)); South Australia – *Occupational Health, Safety and Welfare Act 1986* (OHSWA 1986 (SA)); Tasmania – *Workplace Health and Safety Act 1995* (WHSA 1995 (Tas)); Victoria – *Occupational Health and Safety Act 1985* (OHSA 1985 (Vic)); Western Australia – *Occupational Safety and Health Act 1984* (OSHA 1984 (WA)).

5    The plant regulations and generic plant codes of practice were: Australian Capital Territory – *Occupational Health and Safety Regulations 1989* (OHSR 1989 (ACT)), *National Standard for Plant Approved Code of Practice 1995* (NSP ACOP 1995 (ACT)); Commonwealth – *Occupational Health and Safety (Commonwealth Employment) (National Standards) Regulations 1994* (OHSR (CE) (NS) 1994 (Cth)); New South Wales – *Occupational Health and Safety Regulation 2001*, chapter 5 (OHSR 2001 (NSW)); Northern Territory – *Work Health (Occupational Health and Safety) Regulations 1992* (WHR (OHS) 1992 (NT)); Queensland – *Workplace Health and Safety Regulation 1997* (WHSR 1995 (Qld)), *Plant Advisory Standard 2000* (PAS 2000 (Qld)); South Australia – *Occupational Health, Safety and Welfare Regulations 1995*, Part 3 (OHSWR 1995 (SA)); Tasmania – *Workplace Health and Safety Regulations 1998* (WHSR 1998 (Tas)); Victoria – *Occupational Health and Safety (Plant) Regulations 1995* (OHSR (Plant) 1995 (Vic)), *Approved Code of Practice – Plant 1995* (ACOP–Plant 1995 (Vic)); Western Australia – *Occupational Safety and Health Regulations 1996*, Division 3 (OSHR 1996 (WA)).

referenced in the generic plant codes of practice or given the status of approved codes of practice in their own right (see Johnstone, 2004a, p. 332; Johnstone, Bluff and Clayton, 2012, pp. 466–8). These were principally Australian Standards but also included some instruments issued by other national or international standards bodies (Johnstone, Bluff and Clayton, 2012, pp. 460–1). Such technical standards provide more detailed specifications and procedures aimed at ensuring machinery and other items are fit for purpose and meet minimum levels of health and safety (Productivity Commission, 2006, pp. 6, 14; Wettig, 2002).

*Management of Health and Safety in Machinery Design and Construction*

*Management of risks to health and safety*
Australian OHS law is constitutive in the sense that it attempts to have regulatees establish and implement arrangements for self-regulation so that they act responsibly and comply with their legal obligations as an ongoing state of affairs (Hutter, 2001, pp. 15–16, 77, 301–2; 2011, p. 12; Johnstone and Jones, 2006). In particular, the general duties in the OHS statutes require regulatees to take continuing, proactive and systematic action. They are flexible about the action that regulatees take to comply and are capable of accommodating technological change (Bluff and Gunningham, 2004; Johnstone, Bluff and Clayton, 2012, pp. 180–1).

The general duties in force at the time of data collection required regulatees to ensure that plant was designed and constructed to be safe and without risks to health in end use or, in alternative forms of expression, to protect people from, or prevent exposure to, health and safety hazards or risks arising from plant.[6] These were absolute duties in the sense that regulatees were required to guarantee, secure or make certain the safety of the plant they produced (Creighton and Rozen, 2007, p. 77; Johnstone, 2004a, pp. 275–80). However, the qualifying expressions 'reasonably practicable', 'practicable' or equivalent expressions were incorporated in, or applied in connection with the duties (Bluff and Johnstone, 2005; Johnstone, 2004a, pp. 210–11). In effect the regulatee would need to implement measures to remove or mitigate a risk, unless the burden (in cost, time or inconvenience) of taking particular action was grossly disproportionate to the risk.

As applied in Australian OHS law these qualifying expressions implicitly entail the management of risks. In order to comply with their legal obligations regulatees would need to identify reasonably foreseeable hazards, they would need to assess the magnitude of the risks arising from those hazards, taking into account the state of knowledge about the hazard or risk and ways of removing or mitigating them, and they would need to determine and implement preventive

---

6   OHSA 1989 (ACT) s 42(1); OHS (CE) 1991 (Cth), s 18(1); OHSA 2000 (NSW), s 11(1)(a); WHA 1986 (NT), s 30B(1)(a); WHSA 1995 (Qld), ss 32(1)–(2); OHSWA 1986 (SA), s 24(1); WHSA 1995 (Tas), s 14(1)(a); OHSA 1995 (Vic), 24(1); OHSA 1984 (WA), s 23(1)(a).

measures commensurate with the risks (Bluff and Johnstone, 2005; Cross, et al., 2000; see also HSE, 2001a).

Such a process of risk management was explicitly called for under the plant regulations and/or generic plant codes of practice in most jurisdictions.[7] Regulatees were to identify hazards and address the risks that might arise in different stages of the life cycle of plant, from design to end use; and in different aspects of use such as maintenance, servicing, repair, inspection, adjustment and cleaning of plant. They were to consider intended use and, to some extent, unintended use which might arise through human error or misuse. They were encouraged to use methods for assessment of risks such as visual inspection of plant and the end use environment, auditing, testing, technical or scientific evaluation, analysis of injury and near miss data, discussions with relevant parties, and quantitative hazard analysis. They were to eliminate or minimize (or reduce) risks and, if they incorporated guarding, they were advised when to use different types of physical and interlocked barriers, and presence sensing safeguarding systems.

*Testing and examination*
In addition to managing risks regulatees were required, as part of their statutory general duties, to ensure testing and examination of the plant they designed and constructed, either for the purpose of complying with these duties, or to discover, eliminate and minimize risks.[8] The plant regulations and generic plant codes also included provisions for testing, albeit less far-reaching in scope than the general duties.[9] They called for testing according to any technical standards that incorporated provisions relevant to the testing of plant.

*Design verification, and notification or registration*
While the provisions aimed at constituting risk management, and testing and examination applied to a wide range of machinery and other plant, the regulations imposed additional obligations for certain types of plant (prescribed plant), requiring verification of the design by applying relevant technical standards or engineering principles, and notification or registration of the design with an OHS

---

7  NSP ACOP 1995 (ACT); OHSR (CE) (NS) 1994 (Cth); OHSR 2001 (NSW) ch 5; PAS 2000 (Qld) pt 1, 2, 12; OHSWR 1995 (SA) regs 3.2–3.3; OHSR (Plant) 1995 (Vic) regs 105, 302–305; ACOP–Plant 1995 (Vic), cll 11.1–11.2–12.2; OSHR 1996 (WA) div 3. See also the provisions in the Queensland statute, WHSA 1995 (Qld) s 22(1), 28, 29.

8  OHSA 1989 (ACT) s 32(1)(b); OHSA (CE) 1991 (Cth) s 18(1)(b); WHA 1986 (NT) s 30B(1)(b); WHSA 1995 (Qld) s 32(3); OHSWA 1986 (SA) s 24(1)(c); OHSA 1985 (Vic) s 24(1)(b); OHSA 1984 (WA) s 23(1)(b).

9  NSP–ACOP 1995 (ACT) cll 14(2), 66(2); OHSR (NS) 1994 (Cth) regs 4.05(2), 4.32; OHSR 2001 (NSW) reg 103(3)(b); WHR 1992 (NT) reg 85(2); WHSR 2000 (Qld) reg 15(1)(a), PAS 2000 (Qld) cll 1.11, 2.3; OHSWR 1995 (SA) regs 3.2.6(2), 3.3.2(c); OHSR (Plant) 1995 (Vic) regs 307, 404, ACOP–Plant 1995 (Vic) cll 12.2, 13.2, 13.3; OSHR 1996 (WA) reg 4.24(4), 4.29(c).

regulator.[10] Such prescribed items typically included cranes, pressure equipment, gas cylinders, elevating work platforms, hoists and amusement structures, among others.

*Provision of safety information*

A further set of provisions, also applicable to a wide range of machinery and other plant, concerned safety information. The statutory general duties required regulatees to make information available about their plant, and the regulations and generic codes of practice also addressed information provision.[11] Collectively, the items most commonly required or advised across these different instruments were information about safe use of the plant, and the intended use or purpose for which the plant was designed and tested. The law in some states and territories also required or advised that information should be provided about different aspects of use across the life cycle of plant, the testing or inspection needed for the plant, and emergency procedures. Some provisions called for information about the knowledge, training and skills for those conducting testing or inspection of plant, the testing actually conducted, and information about residual hazards or risks (the dangers) of plant. Less commonly, regulatees were required or advised to provide details of the design or construction of the plant, and restrictions or prohibitions of certain uses.

*Core Elements of Legal Obligations*

At the level of broad principles the Australian OHS statutes, plant regulations and generic plant codes of practice comprised a series of core elements, as set out in Table 2.1 below. These elements could be conceived as a framework for regulatees to proactively and systematically manage health and safety in the design and construction of machinery and other plant, in order to discharge their continuing OHS legal obligations.

The first set of core elements relate to the management of risks to health and safety. They encompass identification of reasonably foreseeable hazards, assessment of the risks arising from these, and implementation of control measures to eliminate or minimize risks, which are commensurate with the risk. These steps are not only required for 'normal use' but for different stages of the life cycle of the plant, different aspects of use (such as maintenance, servicing, repair, inspection,

---

10   NSP–ACOP 1995 (ACT) cll 69–70, sch 1, 2; OHSR (NS) 1994 (Cth) regs 4.51, 4.52, sch 5, 6; OHSR 2001 (NSW) regs 107–109, sch 1; WHR 1992 (NT) regs 88–90, sch 10; WHSR (Qld) r 15–16, Sch 4; OHSWR 1995 (SA) regs 3.4.1, 3.4.2, sch 4; WHSR (Tas) rs 96–99, Sch 1 and 9; OHSR (Plant) 1995 (Vic) reg 1001–1003, sch 2; OSHR 1996 (WA) regs 4.2–4.7, Sch 4.1–4.3.

11   NSP–ACOP 1995 (ACT) cll 11, 15, 20; OHSR (NS) 1994 (Cth) reg 4.06; OHSR 2001 (NSW) reg 105; PAS 2000 (Qld) cll 1.12, 2.4; OHSWR 1995 (SA) regs 3.2.3, 3.2.7; WHR 1992 (NT) regs 84, 85(2), 86(2); OHSR (Plant) 1995 (Vic) regs 308, 405, ACOP–Plant 1995 (Vic) cll 14, 16; OSHR 1996 (WA) regs 4.30, 4.31.

adjustment and cleaning of plant), and for both intended and unintended use of plant (such as human error, mistakes and misuse). The other core elements are testing and examination of plant; design verification and submission of information to the OHS regulator (notification or registration) for specific types of high risk plant; provision of safety information; and the application of technical standards. The latter are recognized through adoption in regulations or approved codes of practice, or approved as codes of practice in their own right. The application of technical standards cuts across all of the other elements as these instruments provide more detailed specifications and procedures for risk management, testing and examination, design verification and provision of safety information.

**Table 2.1     Core elements of legal obligations for the safe design and construction of machinery**

- Management of risks to health and safety, including:
  - identification of hazards
  - assessment of risks
  - implementation of control measures to eliminate or minimize (or reduce) risks
  - consideration of risks in different stages of the life cycle, and different aspects of use
  - consideration of risks arising from intended and unintended use
- Testing and examination
- Design verification, and submission of information to the OHS regulator (notification or registration), for specific types of plant
- Provision of safety information
- Application of technical standards in each of the above

These core elements of Australian OHS law provide a basis for the comparative analysis of the European regulatory regime for machinery safety below. For the Australian firms supplying machinery into Europe this regime was part of the framework of state regulation. The European regime used different terms and approaches but some of the core elements of the Australian regime were also represented in the European regime.

## The European Regulatory Regime

*Overview of the European Regime*

The principal instrument from which the regulation of machinery safety in Europe was derived, at the time of data collection, was the *Machinery Directive* of 1998

(European Commission, 1998a; 1998b; 1998c; 2000a; Boy and Limou, 2003, pp. 10–11). This directive was one of the 'new approach' directives which aim to remove technical barriers to trade in particular products by harmonizing national health and safety provisions applicable to them (European Commission, 2000a, pp. 7–8). The *Machinery Directive* was transposed into the law of member states, as for example with the *Supply of Machinery (Safety) Regulations 1992* (now 2008) in Britain or the machinery regulations under the *Equipment and Product Safety Act 2004* (now 2011) in Germany[12] (European Commission, n.d.; HSE, 1998).

There were further directives for specific types of machinery including simple pressure vessels, pressure equipment and lifts which contained similar elements to the *Machinery Directive* but applied them to specific types of equipment (European Commission, 1995; 1997). There were also directives for particular hazards which might arise from machinery, such as noise, and these directives applied in addition to the *Machinery Directive* (European Commission, 2000b).

This review focuses on the *Machinery Directive* as the primary source of requirements relating to the safety of a wide range of machinery and associated safety components. As such this directive was the European instrument most relevant to the study firms.

*Essential Health and Safety Requirements and Harmonized Standards*

While the management of machinery risks under Australian OHS law involved an open-ended process of identifying hazards, and assessing and controlling risks, the process for managing risks under the European regime was more guided and structured. Central to the *Machinery Directive* were the essential health and safety requirements which set out the health and safety standards to be met for a wide range of hazards and risks, and particular control measures required (European Commission, 1998a, annex 1). The essential health and safety requirements emphasized the need to eliminate or reduce risks. This was to be done for different aspects of use of machinery including adjustment, maintenance, cleaning, unblocking, assembly and dismantling, as well as everyday use and operation. Treatment of unintended use was strong in the European regime as the manufacturer was required to envisage not only the normal use of the machinery but also uses which could reasonably be expected, and machinery must be designed to prevent abnormal use if this would create a risk. Otherwise information must be provided about the ways machinery should not be used.

A manufacturer could choose to comply with relevant harmonized standards in lieu of the corresponding essential health and safety requirements. The harmonized standards are technical standards drawn up by the European Committee for Standardization (CEN) and the European Committee for Electrotechnical Standardization (CENELEC) (European Commission, 2002). If machinery complied with a relevant harmonized standard there was a presumption

---

12    Verordnungen nach § 3 des Geräte – und Produktsicherheitsgesetzes.

of conformity with the essential health and safety requirements covered by that standard (European Commission, 1998, para (2), art 5, cl 2). With more than 400 harmonized standards recognized under this regulatory regime, manufacturers had considerably more technical guidance available to them than under the Australian OHS regulatory regime. There were harmonized standards dealing with basic concepts for design of all machinery (A type), for safety aspects of a range of machinery (B1 type), for safety components or devices used on a wide variety of machinery (B2 type), and for a single type of machinery (C type).

*Self-Assessment of Conformity*

Manufacturers intending to place machinery on the market in Europe, or their authorized representatives in Europe, were required to assess the conformity of machinery. This could be a self-assessment if the machinery was not of a type listed in Annex IV of the *Machinery Directive*. Self-assessment involved identifying the essential health and safety requirements applicable to the particular machinery, ensuring the machinery was designed and constructed to comply with the relevant requirements, and preparing a technical file documenting the measures taken to comply with these requirements and any other prescribed information (European Commission, 1998, art 8, cl 2, annex V, cl 3). The file could be kept at the manufacturer's premises (even if overseas) but must be made available promptly in response to a request from a relevant European regulator. The manufacturer was also required to produce a European Community (EC) declaration of conformity with the essential health and safety requirements, and to affix Certification Europe (CE) marking to the machinery indicating the manufacturer's claim that all relevant requirements had been met (European Commission, 1998, cl 1, annex II).

*Conformity Assessment Involving a Notified Body*

For machinery and safety components listed in Annex IV of the *Machinery Directive* the manufacturer was required to involve a notified body to conduct the conformity assessment. The Annex IV items included some types of saws, presses, devices for lifting people and various other machinery types, as well as some safety components. A notified body is a third party designated by member states and required to meet specified criteria of independence, resources and integrity, as well as technical understanding of the machinery they were appointed to assess, and the relevant essential requirements and harmonized standards (European Commission, 1998, annex VII). Under mutual recognition arrangements, certain recognized bodies in Australia could conduct conformity assessments in lieu of European notified bodies (European Commission, 1998, art 8, cl 2; DTI, 1999, pp. 4, 11; Department of Industry 2014).

   If Annex IV machinery or safety components complied completely with harmonized standards encompassing all essential health and safety requirements relevant to the machinery, the manufacturer could declare conformity of the machinery and send a notified body a copy of the technical file. The notified body might then acknowledge receipt of the file but not examine it; verify the technical file, certify that the harmonized standards had been complied with, and draw up a certificate of adequacy; or carry out an EC type-examination for the machinery. For Annex IV machinery or safety components that did not comply with harmonized standards (or for which there were no relevant standards) the manufacturer was required to submit an example of the machinery for EC type-examination by a notified body. The notified body's role was to examine the machinery and technical file and, if satisfied, issue an EC type-examination certificate or, if not satisfied, refuse to issue a certificate.

*Provision of Safety Information*

In the European regulatory regime, all machinery must be accompanied by safety information in the form of warning signs (or pictograms) and instructions, as set out in the essential health and safety requirements (European Commission 1998, annex 1, cll 1.7.2 and 1.7.4). These requirements covered some of the same information items called for in (some of) the Australian statutory duties, plant regulations or generic plant codes. These were warnings of residual hazards, instructions for correct and safe use including normal and reasonably expected use (foreseeable use), ways the machinery should not be used, and safe use across different aspects of use and life cycle stages. The latter embraced putting the machinery or components into service, use, handling, assembly, dismantling, adjusting, maintaining, servicing and repairing machinery. Additional items in the European regime were information about noise emission and vibration levels (the latter for hand-held or hand-guided machinery), advice about installation and assembly to reduce noise and vibration, training (where necessary), and use of machinery in explosive atmospheres (if relevant). Information was to be provided in one of the European languages as well as the language of the country where the machinery would be used.

**Contrasting Core Elements**

At the level of broad principles the European regulatory regime for machinery safety had some elements in common with Australian OHS law. In order to establish which of the essential health and safety requirements applied to particular machinery it was necessary to identify the hazards of the machinery. Hence identification of a wide range of hazards was common to the European and Australian regimes, as were a process of assessment, elimination or reduction of risks, and consideration of different stages of the life cycle, different aspects of use of machinery, and intended and unintended use (see also European harmonized standards for machinery risk assessment CEN, 1997; 2003a; now CEN, 2010). A process of design verification for designated types of machinery and provision of safety information were also elements of both regimes, as was the application of technical standards.

However, in various ways the European regime was more developed than the arrangements required in order to comply with Australian OHS law. In particular, the *Machinery Directive* facilitated the integration of health and safety in machinery design and construction through the structured and guided process of assessment with reference to detailed health and safety requirements. The European regime provided considerably more technical guidance through multiple harmonized standards dealing with machinery concepts and processes, particular health and safety aspects, safety components and devices and specific types of machinery. There was also an infrastructure of technically competent and independent bodies (the notified bodies) to assess the safety of machinery design and construction.

## Conclusion

This chapter has begun to examine the place of state regulation as one of the elements of interest in understanding and explaining business responses within a holistic and plural conception of business behaviour and compliance (Parker and Nielsen, 2011, p. 5). The Australian and European regulatory regimes for machinery safety provided solid frameworks for managing health and safety risks in machinery design and construction. There were core elements common to both regimes, notwithstanding various differences in the substance and properties of provisions, and the wider range of technical standards underpinning the European regime.

The core elements of the legal obligations for machinery safety provided the conceptual framework in this research, for analysing and evaluating manufacturers' action and performance for machinery safety, and for exploring the factors and processes shaping their responses. This process begins in the next chapter, where manufacturers' performance is examined for three of the core elements: those relating to identification of hazards, implementation of risk control measures, and provision of safety information.

# Chapter 3
# Machinery Manufacturers' Performance for Substantive Safety Outcomes

Occupational health and safety (OHS) is about the prevention of death, injury and illness (Bohle and Quinlan, 2000, ch. 1; ILO, 2001, pp. 3, 18), a goal which is reflected in the Australian and European regulatory regimes in focus in this research (see Chapter 2). The research applied this regulatory goal of prevention as the benchmark for manufacturer performance, and three core elements of manufacturers' legal obligations, which were critical for prevention, provided the conceptual framework for evaluating firm performance. These core elements were identification of hazards, implementation of control measures to eliminate or minimize risks, and provision of safety information. They were framed as substantive safety outcomes since machinery could not be regarded as safe and without risks to health unless manufacturers had comprehensively recognized the hazards of the machinery, eliminated these hazards or incorporated effective control measures to minimize the risks arising from them, and provided safety information to support and reinforce risk control measures.

Manufacturers' performance for these substantive safety outcomes was evaluated (by the author).[1] This chapter presents the findings of this evaluation, showing that while some study firms performed well for all of the substantive safety outcomes (exceptional performers), many only performed well for some outcomes (mediocre performers), and some firms performed poorly across all outcomes (poor performers). Whether readers come to this book from the standpoint of safety, human factors, design, regulation, risk management or otherwise, this is a troubling perspective of firm performance and challenges us to reflect on the reasons why. Crucially then, the evaluation of manufacturers' performance for substantive safety outcomes in this chapter provides the foundations for exploring the factors and processes that shaped firm performance, in the remainder of the book.

Before beginning the discussion of hazard recognition, risk control and safety information it is also important to note that while some firms engaged in risk assessment or risk management processes they did not all do so. Also, more informal processes came into play in all firms, as they identified potential sources of harm, and determined what action to take to address them, in the general course of designing and constructing machinery. It is the outcomes for hazard recognition,

---

1 For additional details see the Appendix.

risk control and safety information that are the focus of this chapter. Processes for assessing and managing risks are examined in Chapter 7.

## Hazard Recognition

*Approach to Evaluating the Scope of Hazard Recognition*

Machinery may be hazardous in many different ways, as recognized in the Australian and European regulatory regimes for machinery safety, and in the safety literature (see for example Brauer, 1994; 2006; European Commission, 1998a, annex 1; NOHSC, 1994; Standards Australia, 1996; 2006a; 2006b; 2014). Hazards may arise from the structure or components of machinery, its function and operation, the power source(s) and the end use environment. There may be structural hazards such as sharp edges, projections or obstructions, or a lack of stability or soundness which may cause machinery to overturn, fragment, collapse or otherwise fail to support or contain its load or contents. There may be mechanical hazards associated with the moving parts of machinery which create the potential for entanglement, crushing, trapping, cutting, stabbing, puncturing, shearing, abrasion, friction, tearing, stretching and generation of projectiles.

There may also be physical hazards including electricity, pressurized content, noise and vibration, ionizing and non-ionizing radiation, heat, cold and moisture (or lack of it). Ergonomic hazards may arise from awkward working positions, manual handling, repetitive movements, static load, poor design of controls or low visibility. Slips, trips and falls may be caused by access and egress hazards associated with poor or absent walkways, stairways, railings, hand-holds or slippery surfaces. Chemical hazards may arise in the form of gases, liquids, vapours, dust or fumes which may cause adverse health effects and hazardous events such as fire and explosion. Other hazards may involve the location and environmental conditions for end use of machinery, such as proximity to people and other machinery or equipment, terrain, temperature, falling objects, services (gas, electricity supply) or impact on workplace design and layout. Biological hazards (for example bacteria or moulds) may be present in materials used or processed in machinery.

These examples are not exhaustive but serve to illustrate the diversity of hazards, and to demonstrate that a great deal more is at stake than mechanical hazards and the need for guarding. For manufacturers to comprehensively recognize hazards, they would need to identify the different types of hazards, and different instances of each hazard for their machinery.

The researcher identified the hazards for particular machinery through observation of machinery at manufacturers' premises or supply outlets, and with reference to technical standards, OHS regulator guidance materials and other published information relevant to the machinery (for example Brauer, 1994;

2006; CEN, 1997; 2003a (now CEN 2010); NOHSC, 1994; 2005; Standards Australia, 1987; 1994a; 1996; 2000a; 2000b; 2000c; 2002; 2006a; 2006b). The hazards actually recognized by each manufacturer were determined on the basis of interview data and other information provided by the firm; for example product brochures, risk assessment reports and machinery safety information. The key individuals interviewed for this research often used terms other than 'hazards' when discussing aspects of their machinery with the potential to cause harm. They referred, for example, to 'safety issues', 'safety problems' or 'safety aspects' (see Chapter 7). Of interest for the research were all the potential sources of harm recognized by firms and not only those actually named as 'hazards', and whether they were controlled in some way or not.

Three levels of performance were determined for hazard recognition. These were: (1) *comprehensive* – the firm recognized the key hazards for its machinery, including the different types and instances of hazards; (2) *incomplete* – the firm recognized some different types but not all types or instances of hazards for its machinery; or (3) *blinkered* – the firm had a narrower conception of machinery safety, focusing on mechanical hazards and overlooking other types of hazards for its machinery. Table 3.1 below presents the proportion of firms with comprehensive, incomplete and blinkered hazard recognition.

*Scope of Hazard Recognition by Manufacturers*

Less than one third of the study firms (30 per cent (20/66)) had comprehensively recognized the key hazards for their machinery, as summarized in Table 3.1 below. Manufacturer 15, which designed and constructed complex processing plant, exemplified this approach. The firm recognized routine mechanical hazards of rotating equipment, electrical hazards associated with faults in or access to live wiring and equipment, structural stability, space for access, lifting associated with manual tasks and entry into confined spaces, including the potential for falls. The firm was also examining emerging hazards associated with a new type of plant as the safety manager explained:

> ... when we start getting into the newer cutting edge stuff ... particularly because we're working with [waste] there is a whole stack of environmental and health problems ... They're using a new process for breeding the bugs that chew up the [waste]. It involves a whole stack of floating things that provide little cubbyholes for the bugs to go and live in ... Now we're suddenly going to be left with tonnes of this media in there that's actually still going to be active and live ... when you get the equivalent of a shipping container load in a very small, confined area that's ... a lot of volume and a lot of weight ... Nobody's got any idea about its stability ... what angle it should be on. Could we have somebody working in the tank and all of a sudden all this stuff all shifts down on him? (Safety manager, Manufacturer 15)

**Table 3.1    Scope of hazard recognition**

| Description | n | % |
|---|---|---|
| Comprehensive: recognized the key hazards for their machinery, including different types and instances | 20 | 30 |
| Incomplete: recognized some different types of hazards for their machinery, but not all types and/or instances | 37 | 56 |
| Blinkered: only recognized mechanical hazards for their machinery and not other types of hazards | 9 | 14 |
| Total | 66 | 100 |

This manufacturer comprehensively recognized a diverse range of hazards for its processing plant. As this example illustrates, these included the health hazard of live biological agents and the safety hazard of a large mass and volume of unstable material similar to a soil excavation hazard.

Many of the study firms had incomplete hazard recognition (56 per cent (37/66)), as summarized in Table 3.1. They recognized different types of hazards, but overlooked some types or instances of hazards for their machinery. Manufacturer 59 was a typical example. This firm customized materials handling systems integrating conveyors, scissor lifts, hoists and semi-robotic applications. The firm recognized mechanical hazards associated with rotating shafts, rollers, belts and nip points, the potential for machinery to tip over, the structural load of overhead systems, and some ergonomic hazards relating to operator posture and controls. The potential for robotic equipment to lose its load was considered with instructions to mark exclusion areas in the end use work environment.

Although this firm considered a range of hazards, they did not recognize all instances of mechanical, structural, electrical and ergonomic hazards for their machinery, and some hazards arising from proximity to people and other machinery in the end use work environment. Of particular concern were robotic elements of the machinery, which had the potential to execute rapid, high energy, unpredictable movements, eject or drop materials handled, and trap body parts between the robot and other fixed objects or parts of the robot (Gunitalaka, 2005; WHSQ, 2006, pp. 33–4). Other hazards with this technology were the potential for control malfunction arising from electrical, hydraulic or pneumatic failures, software faults or other electronic interference.

The remaining study firms had blinkered hazard recognition (14 per cent (9/66)), as summarized in Table 3.1. These firms had a narrow conception of machinery safety, equating it with mechanical hazards, and did not consider other types of hazards for their machinery. Manufacturer 14, which made presses and various types of special purpose machinery, was a paradigm case as the firm focused on the crushing, cutting and shearing hazards if operators accessed dangerous areas of the machinery. This could occur if guards were removed or the machinery was energized while a worker was removing a blockage or during maintenance. These were significant hazards requiring attention but other important hazards were overlooked, including structural, electrical, noise, hydraulic and compressed air hazards (Standards Australia, 1996; WHSQ, 2006, pp. 28, 30). There was also the potential for spills or waste generated by the machinery's processing of materials, and ergonomic issues relating to design of operator controls, lighting and materials handling

Overall more than two thirds of the study firms had not recognized all types of hazards for their machinery, or had overlooked particular instances of hazards. These firms had failed to satisfy the first fundamental step in ensuring that their machinery was designed and constructed to be safe and without risks to health. Their performance for hazard recognition was not enough to support the regulatory goal of prevention. Also, if a manufacturer failed to recognize a particular hazard, that hazard was not even considered for elimination or other control measures to minimize the risk.

**Risk Control**

*Approach to Evaluating the Type and Quality of Risk Control Measures*

There are different types of risk control measures and some are more effective than others in eliminating or minimizing risks. One category of measures is *safe place* controls. These eradicate or reduce the danger by eliminating hazards or incorporating physical safeguards which prevent harmful forms and levels of energy from being generated, released or transferred to people (Adams, 1999; Atherley, 1975; 1978; Haddon, 1973; 1974; 1980; Manuele, 1999b). Such controls include changes to machinery design; substitution of less hazardous components of machinery; isolation to separate people from hazardous areas; and incorporation of engineering controls. The latter extend to interlock mechanisms, guards that are permanently fixed or cannot be opened or removed without a special tool or key, and presence-sensing systems (CEN, 2003b (now CEN 2010); Standards Australia, 1996; 2006c). Safe place controls are more effective because they are less likely to be weakened by human action or error.

The other broad category of ways to reduce risks is *safe person* measures. They require that individuals who operate, maintain or otherwise encounter machinery are continually alert to and actively avoid risks, and take care to protect themselves and others (Atherley, 1975). Examples of safe person measures are warning signs or devices, safe work practices, and personal protective equipment such as gloves, safety glasses and hearing protection. Safe person measures may be rendered ineffective or their efficacy impaired by worker fatigue, human errors or mistakes, and other everyday pressures of work. Consequently, they are less effective than safe place controls and, while they might be used to supplement the latter, they should not be the only risk reduction method used (Dodge, 2001; Manuele, 1999b; see also Kletz, 1991, ch. 1).

In this research a further distinction in the quality of controls was inductively devised through systematic reflection on the data for the risk control measures implemented by manufacturers. Some firms used safe place controls that were more *advanced or innovative*, involving more sophisticated technology or applying technology in original ways that enhanced the safety of the machinery. Advanced or innovative safe place controls were more effective because they used fundamentally different machinery designs to eliminate hazards or minimize risks, or integrated control measures with the design and construction of the machinery so they could not be removed or disarmed. Examples were more sophisticated forms of interlocked guarding systems, mechanisms to de-energize machinery and disperse residual power before guards could be opened, and use of safeguards in series or in combination so as to achieve greater risk reduction.

Other safe place controls were the more *basic or standard* measures used in industry for particular types of machinery. They could be simple but effective if they eliminated hazards by, for example, directly mounting motors or gearboxes onto drive shafts in order to eliminate moving sprockets, chains and belts, or removing sharp edges, corners or protrusions that might cause cuts or bruising. They could reduce ergonomic problems by changing the height of machinery or the reach distance to controls to better suit operators, or reducing the weight of component parts. Also, the machinery might conform to standard recommendations for structural steel, welding, electrical wiring, stairs, platforms and railings (for example CEN, 2001a; CENELEC, 2000; Standards Australia, 1981; 1992; 1994b; 2000d). However, some basic or standard measures were less effective in controlling risks, as with guards that restricted but did not prevent access to the danger zone of machinery while it was operating. Examples of less effective guarding were partial or mesh guards that permitted access through or around the guard; guards that were easily opened, removed or disarmed; and guards requiring manual rather than automatic isolation of power (which could be opened without a worker first isolating the power). Other examples of less effective guarding were interlocked or presence-sensing systems that could be defeated, or were not coupled with a mechanism to dissipate residual power before workers could enter the danger zone (Backstrom and Döös, 1997; 2000).

The researcher sourced information about recommended risk control measures in relevant technical standards, OHS regulators' guidance materials and other published information about the particular type of machinery (for example Brauer, 1994; 2006; CEN, 2003b; Corlett and Clark, 1995; Marras and Karwowski, 2006, pt III, V, VIII; Standards Australia, 1987; 1994a; 1996; 2000a; 2000b; 2000c; 2002). The type and quality of risk control measures actually used by study firms were determined through observation of machinery at their premises or supply outlets, and additional insights were gained through interviews and other information provided by firms such as product brochures, risk assessment reports and machinery safety information. As with conceptions of hazards, the key individuals interviewed often used more everyday terms when discussing the action they had taken to address sources of harm (see Chapter 7). They referred, for example, to 'safety features', 'safety measures' or 'safety devices'. All of these were of interest as potential risk control measures.

*Types of Risk Control Measures Used by Manufacturers*

Just under half the study firms used safe place controls as the primary measures to eliminate or minimize risks with their machinery (47 per cent (31/66)), as summarized in Table 3.2. These firms only used safe person measures to supplement safe place controls, but not instead of them. Manufacturer 21 exemplified this approach. The firm designed and constructed timber processing machinery, in which high speed timber handling conveyors were completely enclosed in tunnels and saws were also fully enclosed. The firm installed slow speed machinery inside two metre high fences, which were fixed at a distance greater than arm's length, in order to prevent anyone from accessing the machinery by reaching over or through the fence. Keyed alike access required that to access the machinery it must be shut down before the key could be removed from the isolation switch, and this key was then required to open the access gate. In addition, the firm used energy dispersers to ensure that when power was isolated the machinery was totally de-energized before anyone could access it, and programmed electronic control systems to prevent manipulation in a way that could compromise safety. Other control measures were variable frequency drives to minimize the need for hazardous hydraulic components, and dust collection systems which were designed to operate to zero dust emission.

Just over half of study firms (53 per cent (35/66)) used a mix of safe place and safe person controls, but relied exclusively on safe person measures for particular risks, as summarized in Table 3.2 (p. 32) A typical example was Manufacturer 48, which designed and constructed agricultural machinery. The firm incorporated safe place controls for some risks, including changing the design of a post-driver so that it would not operate unless a fixed safety cage was closed, thereby preventing access to the danger zone. Spring-loaded catches fitted against the hammer of the post-driver and another catch lower down stopped the hammer from falling in the event that the hydraulic mechanism was damaged or failed. The firm had redesigned a wood-chipping machine to prevent access to cutting blades by lengthening the feed in chute, supplied chains with slashers to reduce projectile hazards and fitted fixed guards on power take-off drives.

For some other risks, this firm relied exclusively on safe person measures. For example, signs on post-hole diggers warned users of the danger of contact with the unguarded spinning auger, and signs on slashers warned users of the risks of working under them while they are running, or without adequate support for the unit. The firm also advised users generally to wear hearing protection to reduce exposure to noise, and eye protection to reduce contact with hydraulic fluids.

**Table 3.2      Type and quality of risk control measures**

| | Quality of risk controls | | | | | |
|---|---|---|---|---|---|---|
| | Advanced/ innovative | | Basic/standard | | Total | |
| Type of risk controls | N | % | n | % | n | % |
| Use safe place controls as primary measures | 4 | 6 | 27 | 41 | 31 | 47 |
| Safe place controls for some risks & safe person only for other risks | 7 | 11 | 28 | 42 | 35 | 53 |
| Total | 11 | 17 | 55 | 83 | 66 | 100 |

Although some firms relied on less effective safe person measures for some risks, all firms made some use of safe place controls. There were, however, differences in the quality of these safe place controls, as set out below.

*Quality of Safe Place Controls Used by Manufacturers*

A small number of study firms (17 per cent (11/66)), as summarized in Table 3.2 above, applied advanced or innovative safe place controls, which involved more sophisticated technology or using technology in original ways. Two paradigm cases were Manufacturer 21 and Manufacturer 48, the producers of timber processing machinery and agricultural machinery respectively, as described above.

Manufacturer 21 had identified innovative and effective risk control measures from state-of-the-art overseas manufacturers, including the tunnels to completely enclosed high speed machinery, the mechanism to fully de-energize and disperse residual power in machinery before guards could be opened, and the zero emission dust control system. Such measures were rare among other study firms.

Manufacturer 48 incorporated several devices in series in the firm's post-driver in order to prevent uncontrolled movement of its hammer, and the post-driver's safety cage guard was integral to the machinery. This design prevented the post-driver from operating if the guard was not in place and was considerably more advanced than the partial guards used by other post-driver manufacturers, which did not prevent access to the danger zone and were easily removed (see also Baker, et al., 2005, p. 1). The redesign of a wood-chipping machine, to lengthen the chute used to feed in materials, was a further example of an integral design solution implemented by this firm.

In contrast, most study firms used basic or standard safe place controls (83 per cent (55/66)), as indicated in Table 3.2 above. These were simple or industry standard measures for the type of machinery. Typical examples were the measures used by Manufacturer 38 in the firm's range of commercial cleaning equipment.

They included interlocked guards for robotic machinery, removable guards over moving parts on manually operated equipment and pressure rated pipes that were also insulated against heat. The firm provided space around larger units to enable lifting by forklifts or cranes, fitted equipment with standard three phase plugs and sockets to enable safe removal without an electrician, and protected electrical components against dust and water ingression. The firm also provided disposable filter bags to customers who considered that washable bags were a health and safety risk, and designed hand-held cleaning wands for balanced, two-handed operation.

A second example of a firm that used basic or standard controls was Manufacturer 8, which designed and constructed various types of winery production plant including grape presses, crushers, conveyors and fermenting vats. The firm's safe place controls included cover guards, which were bolted in place but could be opened, thereby permitting access to danger zones and requiring workers to manually isolate the power for maintenance access. This manufacturer exemplified the weakest combination of type and quality of risk control since, in addition to using basic safe place controls, the firm relied on safe person measures for some risks. One example was a visual and audible alarm to alert workers to keep clear of a rotating press. Another example was instructions for a vessel that was a confined space with a hazardous atmosphere in end use, which directed rescue personnel to cut the vessel wall to free any person that became unconscious while working inside. This procedure was contrary to standard industry practice, which was to fit vessels with entry holes large enough to enable a person entering a contaminated atmosphere to wear air supplied respiratory equipment (Standards Australia, 2001a, pp. 16–17, 19).

Overall, only a small proportion of study firms (6 per cent (4/66)) implemented the most effective combination of risk control strategies, using safe place controls as the primary risk control measures and including some more advanced or innovative control measures among these. The majority of firms employed weaker risk control strategies. They used basic or standard safe place controls, which might reduce but did not eliminate or minimize exposure to hazards, or they also relied on safe person measures that were less effective because they required workers interacting with machinery to be alert to and actively avoid risks. To the extent that firms addressed risks arising in different aspects of use across the life cycle of machinery or arising through unintended use, they also tended to use a mix of basic or standard control measures and safe person measures, as discussed in the next section.

*Control of Risks in Different Aspects of Use of Machinery*

Once machinery is produced it is not only used or operated to perform a particular function, it is also worked on or interacted with in various other ways in the course of installation, maintenance, repairs, cleaning, clearing blockages, transportation and other ancillary activities throughout its life cycle. Risks may arise in any of these activities. The potential for exposure to risks in different aspects of use was recognized in Australian OHS law and the European regulatory regime for machinery safety, both of which required attention to these risks (see Chapter 2). The principle of controlling risks arising across the life cycle of machinery is also central to the specialist body of knowledge for safe design and construction (ASCC, 2006a; Janicik, 1999; NIOSH, 2006; NOHSC, 2002, p. 9).

Almost two thirds of study firms implemented some measures to address risks to persons engaging with machinery in one or more different aspects of its use (62 per cent (41/66)). Most commonly, firms implemented measures for maintenance (58 per cent (38/66)) and cleaning of machinery (18 per cent (12/66)). Firms more rarely implemented measures to protect those installing (6 per cent (4/66)) or repairing machinery (5 per cent (3/66)), or clearing blockages (2 per cent (1/66)). The types of controls included safe place and safe person measures, similar to those that firms applied more generally for controlling risks in the everyday operation of their machinery.

Examples of safe place controls used to reduce risks in maintenance, cleaning and other ancillary activities included mesh guards to enable inspection and cleaning of machinery without removing guards, micro-switches to shut down machinery if guards were opened, space for access around machinery, and providing platforms or ladders with rails for work at heights on machinery. Examples of safe person measures were signs warning against entry or access to danger zones without isolating power, and procedures or instructions about safe work practices for maintenance, cleaning and other ancillary activities.

Manufacturer 26, which designed and constructed fruit processing machinery, gave specific attention to the risks to workers maintaining and cleaning the machinery. The firm's manager outlined various safe place controls implemented to minimize risks:

> Maintenance people and the cleaners seem to be the ones that get caught, particularly cleaners. They're not aware of what the machine does and how it does it ... Most guards that we make would be mesh or open in some way that you can firstly see whether there's a problem in there ... so it minimizes the amount of time a guard will be removed, and because they're made out of mesh you can generally clean through them with compressed air or other ways, hose down.

> ... on the slide doors, if you have to have them open and an exposed area, you would put a micro-switch on there or whatever to prevent that problem and also have the area guarded so they physically can't reach into there. (Manager, Manufacturer 26)

As this example highlights, workers who clean, maintain or perform other ancillary activities with machinery may be at greater risk, or exposed to different risks from workers who operate and are familiar with the machinery. Yet, study firms' attention to risks arising in different aspects of use across the life cycle of machinery was typically piecemeal and ad hoc, if they addressed these issues at all. There was no evidence that firms comprehensively addressed risks in different aspects of use, and those that did address cleaning, maintenance or other ancillary activities used rather basic or standard controls, or safe person measures. Response for this element of the legal obligations and central principle of safe design was generally weak.

*Control of Risks Arising from Unintended Use*

As well as the intended everyday use and operation of machinery, and the different aspects of use outlined above, end users may perform functions or use machinery in ways that are not intended by those designing and constructing it. The real use and operation of machinery often deviates from intended use for a number of reasons, which are well recognized in the safety literature (Backstrom and Döös, 1997; Benedyk and Minister, 1998; Fadier and de la Garza, 2006; 2007; Fadier, de la Garza and Didelot, 2003; Kanis, 1998; Neboit, 2003; Polet, Vanderhaegen and Amalberti, 2003; Reason, 1990, p. 9; 1997, 171–6; Weegels and Kanis, 2000). End users may make unintentional errors (slips, lapses and mistakes) due to poor machinery design and construction, inexperience, fatigue or other work pressures. They may take remedial action in response to machinery faults, or deliberately disarm or remove safeguards that hinder the use of machinery. They may adapt or fail to follow safe working procedures, in order to maintain operations and productivity to the level required by management. They may also use machinery in ways that are different from those anticipated by those designing and constructing machinery simply because it is possible to do so, or they may use machinery differently in order to reduce strain, pain, effort or inconvenience. Whether intentional or inadvertent, these deviations in use may be detrimental to the health and safety of end users, and others working with or near the machinery. Yet such deviations may be tolerated in workplaces, especially if they maintain or increase productivity, limit losses, save time, or reduce the down time for maintenance or clearing of blockages.

Some provisions in Australian OHS law required or advised designers and manufacturers to address risks arising from unintended use of their machinery, and the European regulatory regime required attention to all uses that could reasonably be expected (see Chapter 2). The importance of anticipating uses that are unintended but nonetheless foreseeable is more strongly emphasized in the safety literature, which recommends that designers should take into account end users' experience of real work situations and activities, and design to remove or minimize the potential for deviations that compromise safety (Fadier, de la Garza and Didelot, 2003; Fadier and de la Garza, 2006; Polet, Vanderhaegen and Amalberti, 2003).

For the study firms there were no examples of fundamentally new designs to minimize the potential for risks to arise through unintended use. At best, a small proportion of firms incorporated some safe place controls that coincidentally reduced the potential for at least some types of unintended use with their machinery (15 per cent (10/66)). For example, some firms used more sophisticated interlocking systems that could not be readily removed or disengaged, coupled with de-energizing mechanisms to disperse residual power before guards could be opened. These measures made it difficult to access a dangerous area either deliberately or inadvertently. Similarly manufacturers of pressure vessels prevented over pressurizing of this equipment by incorporating safety valves that could not be isolated by end users.

Other manufacturers, if they addressed unintended use, applied safe person measures such as directions about safe use which they delivered in signs, instructions or other forms of safety information, as discussed further below. On their own such measures did not constitute a reliable and effective response to risks arising from unintended use. This reliance on safe person measures was consistent with the prevalent attitude of key individuals in over half of the firms that end users acted unsafely with machinery, and were at fault or to blame for what these key individuals perceived to be misuse of machinery (55 per cent (36/66)). This perspective suggests a lack of appreciation among the key decision makers in firms of the likelihood that unintended use will arise through poor design, faults or failures in machinery, human error, or response to production and other pressures in the operational setting, as distinct from misuse. As demonstrated further in Chapter 8, firms in which the unsafe worker attitude prevailed were less likely to control risks effectively.

*Some Observations about Risk Control*

Risks can be most effectively eliminated or minimized through safe place controls which may be supplemented, but not replaced, by safe person measures (Atherley, 1975, 1978; Manuele, 1999b; Haddon, 1973, 1980). While all of the study firms used some form of safe place controls, most firms used more basic or standard types, and over half relied exclusively on safe person measures for some risks. In addition, firms focused their risk control action on the routine operation of machinery, paying little attention to maintenance, cleaning or other aspects of use, and only addressing unintended use to the extent that the controls they incorporated for routine operation coincidentally addressed the potential for unintended use, or they provided directions about working safely with the machinery. These findings of weaknesses in risk control resonate with European research documenting the poor design, quality and removability of risk control measures (Boy and Limou, 2003, pp. 61–4).

Overall, many of the study firms did not produce machinery that was inherently safe. While it was not essential for firms to use more advanced or innovative risk controls, those that used weaker types of basic or standard controls, and those that relied exclusively on safe person measures for some risks, had not ensured that their machinery was safe and without risks to health (so far as reasonably practicable). Their performance in regard to the type and quality of their risk control measures was not sufficient to sustain the regulatory goal of prevention.

**Provision of Safety Information**

*Approach to Evaluating the Scope and Quality of Safety Information*

Risks are most effectively eliminated or minimized through safe place controls, but provision of safety information has an important supplementary role to play in supporting and reinforcing these control measures. Information can advise customers and end users about risk control measures and the action they need to take to ensure the integrity of risk controls. It can alert customers and end users to residual hazards and additional precautions needed to work safely with machinery in different aspects of its use. Standard setting theory also suggests that informed customers can influence manufacturers to control risks in order to avoid loss of sales, a form of market-based regulation (Morgan and Yeung, 2007, pp. 96–7).

To be effective in supporting preventive action or purchasing decisions by customers and end users, machinery safety information must be comprehensive in scope so that it is informative, and of good quality so that it can be easily used and understood. In this research, the scope of manufacturers' machinery safety information was evaluated with respect to its content and form, and quality with regard to the presentation and structure of the information.

The researcher examined the content of machinery safety information to determine whether the manufacturer provided information about hazards and risks, safety features and safe use for different aspects of use, emergency procedures, intended use, restrictions or prohibitions of certain uses, and testing or inspection required. These were types of information that manufacturers were required or advised to provide under Australian OHS law and the European regulatory regime for machinery safety, (see Chapter 2).

In evaluating the form of safety information the researcher considered the different methods used by manufacturers to deliver information. Methods including manuals, instructions, labels, warnings, decals or markings were variously suggested in some of the Australian plant codes of practice and technical standards, any of which might incorporate pictograms; that is, standardized picture safety symbols (see for example Standards Australia 1987; 1994a; 1996; 2000a; 2000b; 2000c; 2002). The European *Machinery Directive* was more explicit, requiring information in the form of warning signs or pictograms, as well as instructions (European Commission, 1998a; annex 1, cll 1.7.2, 1.7.4; see also CEN, 2010).

Taking into consideration both the content and form of machinery safety information, the researcher assessed the scope of the information as being *substantial* if it included more comprehensive warnings of residual hazards or risks, detailed advice about safety features and safe work practices for the machinery, and this information was provided in a combination of labels, instructions, manuals or other forms. Safety information was assessed as *little* if the manufacturer provided a small amount of information about machinery safety in a label, instructions, manual or another form. A third category, *negligible or none* was distinguished if the manufacturer either provided very little safety information in a single sign or a few lines in an operator manual or instructions, or there was no evidence the firm provided safety information to customers or end users, and there were no signs or markings visible on the machinery.

For machinery safety information that was substantial or at least little in scope (but not negligible or none), the researcher also evaluated its structure and presentation as indicators of the quality of that information. *Good* quality information was easy to locate, read and understand; and *poor* quality information was hard to locate, read or understand.

This assessment was informed by the safety literature and relevant technical standards which advise that for operator manuals and instructions to be located, read and understood they should have a list of contents and index, definitions for key terms, logical sequencing of information, numbered headings, and good layout and spacing (Konz, 2006; Reunanen, 1993, app 1; Standards Australia, 1996; and see CEN, 2001b; 2003b (now 2010)). They should also present information in concise statements in active voice, as step-by-step procedures, avoid cross-referencing and use illustrations linked with the text. For labels and markings to be clearly visible and understandable, recommended features are the use of large print, simple statements in plain language, signal words (danger, warning), pictograms, contrasting colours, placement of labels and markings where they are likely to be seen, and durability to resist damage (Brauer, 2006, p. 73; Corlett and Clark, 1995, pp. 87–8, 113; Standards Australia, 1994c). In all forms of safety information the term 'danger' should be reserved for situations that might be life threatening and the term 'warning' for hazardous but not life threatening situations; and the colour red should be used to signal danger, yellow to warn of hazards, blue for required safety actions, and green for emergency exits and equipment (Standards Australia, 1994c; see also ISO, 2013).

The evaluation of the scope (content and form) and quality (structure and presentation) of machinery safety information which follows is based on interviews with manufacturers, their information materials, and any signs or markings on their machinery. Table 3.3 below presents the proportion of firms providing safety information by the scope and quality of information.

*Scope of Machinery Safety Information*

Of the content items prescribed or advised in Australian OHS law and the European *Machinery Directive*, as set out above, study firms most commonly provided advice about safe practices or personal protective equipment for work with machinery (68 per cent (44/66)), and warnings about residual hazards or risks (65 per cent (43/66)). Firms' information did not address all aspects of use across the life cycle of machinery as they focused on the routine operation of machinery and, at most, some provided information about installation, maintenance or servicing. Firms rarely provided information about emergency procedures, testing and inspection, or the intended use and restrictions on the use of machinery. They did not provide information about the training and skills required for persons testing and inspecting machinery other than rarely, and only briefly, mentioning the need to use qualified electricians for electrical work.

These findings about weaknesses in the content of manufacturers' safety information resonate with findings from a European study on the implementation of the *Machinery Directive* in relation to wood working machinery (Boy and Limou, 2003, pp. 37–9). This study found that machinery instructions often contained insufficient safety information; that this information was added to general operator instructions; and that the information did not properly address different aspects of use across the life cycle of machinery, correct machine operation, and testing or inspection for faults.

Among study firms, the more common modes of delivery were a general operator manual that included some safety information (55 per cent (36/66)), and decals, placards or other markings on the machinery (50 per cent (33/66)). Some firms provided one or two-page operator instructions or procedures (17 per cent (11/66)), and some provided a risk assessment report as the safety information, which they added to the operator manual or provided separately (17 per cent (11/66)). Less commonly firms recognized that customers or end users might not read or understand written information, and produced audio–visual materials (6 per cent (4/66)), provided training (9 per cent (6/66)) or one-on-one explanation (14 per cent (9/66)), or arranged for distributors to provide this explanation (3 per cent (2/66)). Like operator manuals, these alternative modes delivered safety information within general information about the machinery.

The various forms of machinery safety information were part of the repertoire of safe person measures used by manufacturers (see above). Firms used such measures to warn customers and end users of residual hazards or risks of machinery, and to provide advice about safe work practices or personal protective equipment, so that customers and end users might then take steps to protect themselves and others. A key reason for providing information was simply that a manufacturer recognized a particular hazard and determined the need to provide information as part of the firm's approach to risk reduction. Firms adopted the format of instructions or manuals, labelling or marking because these methods reflected common practice among equipment manufacturers more generally in providing information to customers (see also Standards Australia 2008a; 2008b). Rather than giving careful thought to how to provide safety information effectively, many manufacturers simply added a little safety information to their general operator instructions, manuals, labels or markings.

Considering both the content and form of safety information, less than one third of study firms provided information that was substantial in scope (27 per cent (18/66)), as summarized in Table 3.3 below. They provided comprehensive warnings of residual hazards or risks, and detailed advice about safety features

**Table 3.3     Scope and quality of machinery safety information**

| Scope (content & form) | Quality (presentation and structure) | | | | | | | |
| --- | --- | --- | --- | --- | --- | --- | --- | --- |
| | Good | | Poor | | Negligible/none | | Total | |
| | $n$ | % | $n$ | % | $n$ | % | $n$ | % |
| Substantial | 16 | 24 | 2 | 3 | – | – | 18 | 27 |
| Little | 15 | 23 | 13 | 20 | – | – | 28 | 43 |
| Negligible/none | – | – | – | – | 20 | 30 | 20 | 30 |
| Total | 31 | 47 | 15 | 23 | 20 | 30 | 66 | 100 |

and safe work practices for the machinery, in a combination of operator manuals, markings on machinery, instructions or procedures. Some of these firms reinforced their written information with audio–visual materials, training or explanation to customers and end users.

The provision of substantial information was exemplified by Manufacturer 29, a producer of agricultural machinery, which fixed decals to the machinery and provided an operator manual. The manual included a separate chapter of safety information, and integrated warnings and safe work practices in the sections of the manual providing advice about assembly, installation, operation, transport, lubrication and maintenance of the machinery. The firm also had an arrangement with its distributors to explain the operator manual to customers at the time of supply.

However, as summarized in Table 3.3, a large proportion of study firms provided little safety information (43 per cent (28/66)), which they presented in labels or markings, brief instructions or a short section within a general operator manual. Manufacturer 40, which designed and constructed complex industrial furnaces, typified this approach in providing only five pages of safety information within a lengthy operator manual, and no additional labels or markings on the machinery.

The remaining study firms provided negligible or no safety information (30 per cent (20/66)), as summarized in Table 3.3. They either provided a single sign or a few lines in an operator manual or instructions; or there was no evidence that they made any safety information available to customers and end users. An example was Manufacturer 2, which produced a drilling rig on which there was a single sign indicating its safe working load.

Overall the performance of the majority of manufacturers in relation to the scope of their safety information was poor as they provided only a little, negligible or no safety information (73 per cent (48/66)). There was no evidence that these firms determined what information their customers or end users might need to protect themselves and others, or what information they should provide in order to comply with their legal obligations. The limited scope of these firms' safety information was incapable of facilitating market pressure to improve the safety of machinery, and nor was there any evidence that firms experienced market pressure to improve the scope of their safety information. The consistency between the findings of this and European research on the implementation of the *Machinery Directive* (Boy and Limou, 2003, pp. 37–9) suggests that while providing some form of machinery safety information is common industry practice, there is a wider problem that manufacturers do not ensure their information is comprehensive and reinforced through various modes of delivery. Weaknesses in the scope of safety information were also exacerbated by the poor quality of some firms' information.

*Quality of Machinery Safety Information*

This research revealed a series of weaknesses in the structure and presentation of manuals and instructions. These ranged from lack of formatting, poor organization or sequencing of information and small print size to poor, wordy or unnecessarily technical expression. Information might also be hard to find as it was a small section (or sections) within long instructions or manuals, or essential details were contained in cross-referenced sources. Labels and other markings used to convey safety information also had weaknesses including dense wording and small print or pictograms, which were hard to read or not clearly visible.

Some firms used terms, colours or pictograms in non-standardized ways, thereby reducing the impact and efficacy of their warnings and directions (European Commission, 1992; Standards Australia, 1994c, pp. 5–6, 17). Examples were use of the term 'danger', which should be reserved for life threatening hazards, to indicate less serious hazards, and use of the term 'caution' to indicate a hazard, rather than the standard term 'warning'. Also, some firms used colours that were different from the standard combinations; for example, using yellow instead or red to indicate life threatening danger.

Taking both structure and presentation into consideration, about half the study firms provided good quality safety information (47 per cent (31/66)), as summarized in Table 3.3 above. Their safety information in labels, markings, manuals or instructions was easy to locate, read and understand. A case in point was Manufacturer 29, the agricultural machinery manufacturer that also provided substantial information, as described above. The firm fixed conspicuous and easy to read decals to the machinery, and the operator manual was designed to be user-friendly. Key information in the manual was highlighted with appropriate use of the signal word 'warning', and concise instructions were coupled with illustrations to indicate prohibited, unsafe and correct practices. Other features enhancing accessibility of information in the manual were a table of contents, definitions for key terms, sequencing of information by different aspects of use of machinery, easy to read print size, simple expression and well-spaced layout.

On the other hand, as summarized in Table 3.3 above, almost one quarter of study firms provided safety information that was poor in quality as the information was hard to locate, read or understand (23 per cent (15/66)). A typical example was the operator manual provided by Manufacturer 62, a producer of agricultural machinery. Critical information about the machinery's potential to cause fatal injuries and safe work practices to reduce this risk was buried in verbose description, in small print, about other aspects of machinery operation. In another example the operator manual provided by a manufacturer of metal cutting and processing machinery (Manufacturer 55) required specific expertise to understand as it contained highly technical safety information. The remaining firms' information was not evaluated for quality as it was either negligible or non-existent (30 per cent (20/66)).

*Some Observations about Machinery Safety Information*

While more than two thirds of study firms provided some machinery safety information (70 per cent (46/66)), their information differed in scope and quality. About one quarter of firms provided substantial, good quality safety information (24 per cent (16/66)), but almost half provided information that was limited in content and form, and/or inaccessible or hard to understand (46 per cent (30/66)). The remaining firms provided negligible or no safety information (30 per cent (20/66)). If not absent, machinery safety information was often poorly equipped to enable customers and end users to make informed decisions about protecting themselves and others, or to create market pressure from customers or end users for manufacturers to improve machinery safety. It is also likely, in many instances, that the information provided would not support the regulatory goal of prevention.

**Manufacturers' Performance in Aggregate**

Taking each substantive outcome in turn, this chapter has shown that 30 per cent of study firms had comprehensively recognized the hazards for their machinery, 47 per cent used more effective safe place controls as the primary measures to eliminate or minimize risks with their machinery, and 24 per cent provided substantial, good quality safety information. On the other hand, 56 per cent of firms had incomplete hazard recognition, 14 per cent were blinkered in their approach, 53 per cent used some safe place controls but relied on less effective safe person measures for some risks, and 76 per cent provided little, negligible and/or poor quality safety information. However, in order to comply with the regulatory goal of preventing death, injury and illness manufacturers needed to perform well for all of the substantive safety outcomes. How then did study firms perform overall?

Reflecting on each firm's performance across all of the substantive safety outcomes this research distinguished three levels of performance – exceptional, mediocre and poor performance. Exceptional performers comprehensively recognized hazards, used safe place controls as the primary risk control measures, and provided substantial, good quality information. That is, they performed well across all the substantive outcomes. Less than one in ten firms (9 per cent (6/66)), were exceptional performers and these were the only ones that could be assessed as having substantively complied with the regulatory goal of prevention. Even among this group, good performance did not necessarily extend to addressing risks for different aspects of use across the life cycle of machinery, using more advanced or innovative safe place controls, and taking action to address unintended use of machinery.

The majority of firms were mediocre performers in the sense that they performed well for one or more, but not all, of the substantive safety outcomes (80 per cent (53/66)). There were weaknesses in their hazard recognition, approach to risk control, and/or the scope and quality of their safety information. There was, however, another group of firms that were especially poor performers (11 per cent (7/66)). They had blinkered hazard recognition, used only basic or standard safe place controls, relied on safe person measures for some risks, and provided little or poor quality safety information, or negligible information. Neither the mediocre nor the poor performers complied with the regulatory goal of prevention. They had not designed and constructed their machinery to be safe and without risks to health, or they had not provided sufficient safety information for their machinery.

These empirical findings of generally mediocre and sometimes poor performance by manufacturers are significant in the context of the long-standing Australian and European legal obligations for machinery safety, and the national and international policy and professional pressures to make items inherently safer through design (Kletz 1998a; 1998b; Manuele, 1999a; NIOSH, 2006; NOHSC, 2002; Swuste, 1997). They indicate that the combined legal, policy and professional pressures have not substantially advanced the goal of preventing death, injury and illness through the safe design and construction of machinery.

How then were manufacturers' responses and performance on machinery safety matters shaped? Why were a small number of firms exceptional performers, while many were mediocre or especially poor performers? As well as evaluating manufacturers' performance for substantive safety outcomes, this chapter has laid the foundations for exploring, in the remainder of the book, the factors and processes that shaped firms' responses on machinery safety matters. The qualitative analysis applied in this research was the means to distinguish the key factors and processes, within and external to firms, which were linked with better or poorer performance.

## Conclusion

This chapter has provided empirical evidence of manufacturers' performance for substantive safety outcomes for hazard recognition, risk control (including for different aspects of use and unintended use), and provision of safety information. This evidence demonstrates that manufacturers' performance was mixed, but that overall the majority of firms did not perform well for one or more of the substantive outcomes. Only a small proportion of firms had comprehensively recognized hazards, used safe place controls as the primary risk control measures, and provided substantial, good quality safety information. The other firms had not ensured that their machinery was designed and constructed to be safe, or had not effectively informed their customers and end users about the safety aspects of their machinery.

The book now turns to explaining these findings for manufacturer performance. It examines the role and influence of state regulation (legislation and enforcement), organizational and individual characteristics and capacities, a wider web of non-state institutions and actors, and the extent to which each of these shaped business goals or priorities, knowledge about machinery safety in firms and, in turn, performance for substantive safety outcomes. In so doing the book illuminates, for machinery manufacturers, key conceptual themes which account for differences in business behaviour, consistent with a holistic and plural conception of compliance (Parker and Nielsen, 2011, ch. 1). The first step in this process is to examine, in the next chapter, manufacturers' understanding of their legal obligations for safe design and construction of machinery.

Chapter 4

# Awareness of Legal Obligations for Machinery Safety

State regulation in the form of legal instruments is one of the elements of interest in understanding and explaining manufacturers' responses on machinery safety matters, as legal instruments may contribute to the knowledge base and motivations for firms to take preventive action. The related role of regulators in supporting, inspecting and enforcing compliance is also crucial, and is examined in Chapter 5.

We saw in Chapter 2 that the Australian and European regulatory regimes provided solid legal frameworks for managing safety matters in machinery design and construction. In principle, manufacturers could apply these in order to proactively and systematically control risks. The present chapter shifts attention from this law in the books to awareness of the relevant legal obligations in study firms. In so doing, the chapter begins to explain firms' performance for substantive safety outcomes, as examined in Chapter 3.

It was reasonable to expect that the Australian and European legal obligations would show up among the influences shaping manufacturers' knowledge and motivations for addressing machinery safety matters, since both regulatory regimes had been in place for many years. In Australia, the occupational health and safety (OHS) statutes in most states incorporated general duties for designers and/or manufacturers of machinery and other plant from the 1980s, and in all Australian jurisdictions from the early 1990s, and these duties were underpinned by regulations and generic codes of practice for plant from the mid-1990s (Johnstone, 1997, pp. 260–3; 2004a, pp. 275–80; see also Chapter 2). In Europe the centrepiece of the regulatory regime for machinery safety, the *Machinery Directive*, was first adopted in the late 1980s (European Commission, 1989).

The lens for examining awareness of legal obligations for machinery safety in study firms was the principal decision maker for design and construction matters (the key individuals). This chapter presents empirical evidence of the generally low awareness of these legal obligations among the key individuals, although the law had a somewhat wider influence as a motivation for preventive action, which was attributable to the perceived threat or authority of the law.

Readers coming to this book from diverse standpoints may react to the findings in this chapter in different ways. For policy makers and regulators who expend considerable effort crafting legal instruments and authoritative guidance it may be disturbing that there is so little engagement with legal obligations in practice. Socio-legal scholars well versed in organizational responses to legal processes may be less surprised about the limited uptake of state regulation. Specialists in human factors or safety engineering, interested in the scientific and technical aspects of risks and their control, may understand the lack of appeal of legal instruments.

For all readers, this chapter adds to our understanding of how safety in machinery design and construction works in practice, and the place of state regulation in this. It introduces the concern that we cannot simply explain low awareness of legal obligations as the product of characteristics such as firm size (James, et al., 2004; Lamm and Walters, 2004; Walters, 2001; and see also Hutter, 2011, p. 147), or the substance and properties of legal obligations (Black, 1997, pp. 22–3; Diver, 1983; Baldwin, 1995, pp. 7–11; and see Cowley, Culvenor and Knowles, 2000). In study firms, low awareness of legal obligations was a much more basic problem. As foreshadowed in this chapter, it had its roots in how those involved in machinery design and construction learned about safety matters through practice, findings which may resonate for readers interested in professional or vocational learning. To the extent that key individuals in manufacturing firms learned something about the law, it was through supply chain and other industry sources that they encountered and drew upon in everyday activities. On the other hand technical standards were a preferred source, whether or not they had any legal standing, as they provided specific information and indicated accepted industry practice.

**Awareness of Australian Obligations**

There was little evidence of awareness among the key individuals in study firms of specific provisions in Australian OHS laws for the safe design and construction of machinery. There were only three firms in which the key individuals had some detailed knowledge of relevant provisions (5 per cent (3/66)). These key individuals were all responsible for safety management in their firms, made use of OHS law in the course of their professional roles, and had ensured that their firms' engineers were trained in the requirements for machinery design and construction. The key individuals were aware of the elements of the legislative framework (OHS Act, regulations and approved codes of practice), the risk management process and the legal force of some technical standards. For example, in one firm the business development manager, who was also responsible for safety matters, had prepared a risk assessment procedure for machinery produced by the business. The procedure applied the relevant regulations, explained the reasonably practicable standard for determining the level of risk control, outlined the type of guarding to be used in different situations, set out arrangements for inspection and testing of the machinery, and referenced applicable technical standards.

In one third of firms the key individuals knew a little about some relevant provisions in Australian OHS law (33 per cent (22/66)). They had learned something about the law through their interactions with the firm's customers, distributors or other industry sources. These individuals were able to say something accurate about the provisions for machinery design and construction, although they might also have misconceptions. For example, an engineer with one firm was aware that different parties had responsibilities under OHS law, and that these were set out in an OHS Act, regulations and code of practice. He did not know the names of these instruments and mistakenly believed that the manufacturer could transfer responsibility for machinery safety to the customer (the employer in end use), rather than the manufacturer and employer sharing concurrent responsibilities, as required by the relevant OHS Act.

Collectively, among the key individuals that knew some detail or at least a little about Australian OHS law, the provisions they knew something about were the general duties (14 per cent (9/66)), design verification requirements (12 per cent (8/66)), and the risk management requirements (33 per cent (22/66)). Also, as discussed in Chapter 7, there were 39 firms that conducted, or engaged a consultant to conduct, some form of risk assessment (59 per cent (39/66)). This was more than the firms in which key individuals were aware of the risk management requirements, as these 39 firms conducted assessments in response variously to European requirements, customer expectations or, more rarely, directions from an Australian OHS inspector. Awareness of particular provisions in Australian OHS law was therefore not a pre-requisite for taking action to assess and manage risks.

In the remaining firms the key individuals knew nothing about the machinery design and construction provisions in the Australian OHS statutes, regulations or generic codes of practice for machinery and other plant (as distinct from technical standards), or they knew only that some OHS law existed (62 per cent (41/66)). They had never accessed or referred to the relevant provisions in the OHS statutes, regulations or codes, even though these instruments were readily available on the websites of the OHS regulators (at no cost) or from government outlets (at low cost). A sample of comments from some key individuals illustrates their lack of engagement with Australian OHS law:

> … we haven't gone out actively and read up on regulations for occupational health and safety. (Director, Manufacturer 2)

> I haven't read that particular regulation and we really are just trying to comply with Australian Standards. (Engineering manager, Manufacturer 9)

> Where do I get that from? Do I ask for it or do they send it to me? (Owner, Manufacturer 24)

> Well, you've got to make it as safe as possible. What the actual law expects I don't know. (Managing director, Manufacturer 28)

> I know that's the *Occupational Health, Safety and Welfare Act* and we would have to do something with the manufacturing – make sure it's safe to use … what it really focuses on with us I wouldn't have a clue. (Manager, Manufacturer 33)

> ... have I read any excerpts from the Act? No. My drivers are something different ... am I aware of it? No. (Manufacturer 39)

> None of us have a copy. I don't know where you would start with the law there. (Managing director, Manufacturer 40)

> I haven't read the Act, I'm not sure that anybody that I know has either. (Managing director, Manufacturer 54)

The generally low level of awareness of Australian OHS law was the same for manufacturers in both study states, South Australia and Victoria. That is, key individuals in a similar proportion of firms in each state knew little or nothing about the machinery design and construction provisions in Australian OHS law.

Rather than OHS Acts, regulations or the generic codes of practice, those involved in machinery design and construction were more likely to refer to and apply technical standards, such as Australian Standards, that were relevant to the machinery they produced (76 per cent (50/66)). Some perceived technical standards to be legal requirements even where they were not called up in regulations, and did not have evidentiary status as approved codes of practice, a misconception shared with industry users of technical standards more generally (Productivity Commission, 2006, p. XIX; and on the status of technical standards see Chapter 2).

Regardless of any perceived legal status, technical standards were a preferred information source for machinery manufacturers because they provided specific information and an indication of accepted industry practice. Referring to technical standards was embraced as part of design and construction practice in a way that, with few exceptions, engaging with the relevant provisions in the OHS Acts, regulations or the generic codes of practice was not.

**Awareness of European Obligations**

The key individuals in a small proportion of study firms were aware that there were European requirements for machinery safety (17 per cent (11/66)). These 11 firms were among the 19 that exported their machinery to countries in Europe or had plans to do so. In three of the 11 firms, the key individuals were aware of some specific details of the European regime as the firms conducted their own conformity assessments for the European market and had obtained information about the European requirements. For example, the director of a firm that manufactured surface finishing machines had tracked information from several sources. She explained:

> ... we checked out all the derivatives [sic] and so forth and we comply with three major derivatives and one is the safety of machinery, one is the low electromagnetic voltage ... .

> ... once we knew that we were interested in exporting I started enquiring about it ... the Business Centre gave me particular names. I went to the web ... I ended up talking to a lady in Canberra ... and she helped me immensely and Austrade helped me also. They initially sent me through some derivatives [sic]. I read them ... it was a long paperwork trail to get there ... but when I got to the actual nitty gritty of what was required it was fairly simple after that, you know complying with that, but just getting to that point was difficult. (Director, Manufacturer 10)

This manufacturer had conducted a conformity assessment, prepared a technical file and affixed CE marking to the firm's machinery in accordance with the European requirements. In eight other firms the key individuals were aware of the European regulatory regime but had little knowledge of specific requirements, as the following comments illustrate:

> ... we have to sign off on it and whatever it calls for we have to meet. We're liable if it sells into Europe without those requirements ... (Managing director, Manufacturer 5)

> ... if a machine comes through to the port of Marseilles and it doesn't have a CE mark it could be a problem ... [the French agent] said providing it's got a little thing on there with CE written on it, it'll pass ... . (Owner, Manufacturer 19)

> ... I'm not absolutely sure of the exact process but it's with an individual testing authority which is recognized ... they sign off on the fact that it meets some particular standard or some level of operational requirements. (Managing director, Manufacturer 57)

> There's an EU agent in Melbourne and we just rang him up, he came out, had a look at everything we made ... and then he went back, wrote a report on how our standards would fair overseas, what we need to improve and what would be required to meet European standards ... (Designer, Manufacturer 64)

Of the eight firms in which the key individuals were aware of the European regulatory regime but knew little about specific requirements, six used consultants who specialized in European conformity assessments and advised them of relevant harmonized standards. The other two (of the eight) firms had not conducted conformity assessments, or used a consultant to do so, and were only aware of particular requirements relating to labelling or electrical safety.

Apart from the three firms that conducted their own conformity assessments and knew some details of the European regime, none of the key individuals in the other firms exporting machinery to Europe actively informed themselves about the process for assessing the conformity of their machinery for the European market.

As with Australian OHS law, there was little evidence of key individuals engaging with the European requirements for machinery safety.

### Explaining the Low Awareness of Legal Obligations

The findings of generally low awareness, or lack of awareness, of Australian and European legal obligations for the safe design and construction of machinery resonate with a series of British studies of safety-related, constitutive regulation. For example, in a study of the large British Rail organization, most of the 120 staff interviewed were aware they had legal responsibilities but were unable to give substantive replies about the nature of this responsibility (Hutter, 2001, pp. 86–7, 96). A study with managers at 40 industrial and agricultural sites, of different sizes, also revealed that ignorance of OHS law was common (Genn, 1993). Similarly, separate studies of agricultural machinery designers, business responses to six sets of UK OHS regulations, and food safety regulation found low awareness of these respective requirements (Crabb, 2000, p. 26; Hanson, et al., 1998; Fairman and Yapp, 2005a; Hutter, 2011, pp. 74–5, 87).

Elsewhere socio-legal scholars have highlighted the design of regulatory provisions as a key issue in compliance, influencing regulatees' interpretation and application of the law, and/or the perceived legitimacy of the law (Parker and Nielsen, 2011, p. 5; see also Haines, 2011; Johnstone, Bluff and Clayton, 2012, pp. 179–82). Regulatory design embraces the substance of legal obligations, as well as their precision (Diver, 1983) and properties (Baldwin, 1995, pp. 7–11; Black, 1997, pp. 22–3). Scholarly analyses have discussed the merits of a precise style of drafting to avoid vagueness, the use of words that are well defined and readily understood within the regulated community, and the framing of regulatory demands so that regulatees can readily determine how to apply them to concrete situations (Baldwin, 1995, pp. 9, 11; Black, 1997, pp. 22–3; Diver, 1983; see also Scott, 2010, pp. 108–11).

Among manufacturing firms the substance or properties of the law did not, in and of themselves, account for low awareness of legal obligations. For most firms, and the key individuals in them, there was a more fundamental problem of failure to engage with the relevant legal obligations, or authoritative sources of information about those obligations. The six firms in which the key individuals knew some detail about Australian OHS law (three firms), or the European regulatory regime for machinery safety (three other firms), were the only ones that made an effort to understand their legal obligations. In the other firms seeking out information about applicable legal obligations was not part of design and construction practice.

Braithwaite and colleagues have proposed motivational posturing theory[1] as an explanation of how regulatees position themselves in relation to a regulatory

---

1   Motivational postures are conscious expressions of underlying motives, beliefs, values, attitudes, interests, goals and feelings towards a regulatory system (Braithwaite V

system (including the law and enforcement by the regulator), so that they do not understand or hear the system's demands and do not fear the consequences of non-compliance (Braithwaite V, 1995, 2009, pp. 74–82; Braithwaite V, et al., 1994; Braithwaite, Murphy and Reinhart, 2007). In manufacturing firms, many of the key individuals displayed features of the posture of disengagement as they did not understand the regulatory system's expectations of them and were not about to ask what their legal obligations were.

It was among small businesses that there was the highest proportion of firms with key individuals lacking knowledge of their legal obligations for machinery safety. In most of these firms the key individuals lacked knowledge of Australian OHS law (82 per cent (28/34)). This compared with half of the medium sized firms (52 per cent (12/24)), and about one third of the large firms (37 per cent (3/8)), in which key individuals lacked knowledge of OHS law.

These findings reinforce other research which suggests that firm size shapes knowledge about safety matters. For example, James, et al. (2004), Lamm and Walters (2004) and Walters (2001, pp. 32, 51, 140, 153) argue that safety knowledge falls with organization size and is linked to the lower presence of or access to safety specialists, lower managerial specialization and systems for management in smaller enterprises, and reliance on non-state, third parties for information about safety and other matters. Smaller firms may also be less visible to OHS inspectors and local communities, and as a consequence less motivated to seek information about OHS requirements (Genn 1993).

Firm size may shape knowledge about safety matters through limits in expertise, information and resources, but among study firms their size alone did not explain differences in awareness of legal obligations for machinery safety. For both the Australian and European regulatory regimes there were small, medium and large firms in which key individuals engaged directly with particular legal obligations and knew some detail about them. There were also examples across the size spectrum of firms in which key individuals did not engage with their legal obligations, or authoritative sources of information about them, and lacked knowledge of how these obligations impacted on machinery design and construction.

Learning about legal obligations for machinery safety was strongest when key individuals engaged directly with particular legal instruments. This was the case for the six firms in which key individuals had some detailed knowledge about Australian OHS law or the European regulatory regime for machinery safety. Learning was weaker, or prone to misconceptions, when key individuals were only aware of some aspect of the legal obligations through their interactions with customers, distributors or other industry sources. Yet learning through everyday interactions was more typical practice and, to the extent that particular legal instruments (Act, regulation or code) or authoritative regulator guidance were part of the mix, they had to compete with more informal, everyday sources (see Chapter 6). And, learning about legal obligations was lacking completely when

---

1995, 2009; Braithwaite, Murphy and Reinhart, 2007).

key individuals did not engage directly with particular legal instruments, had not encountered information about their obligations through their interactions with external actors, and did not have past experience of them from other work settings.

There is some debate about whether specific knowledge of legal obligations is an essential pre-requisite for compliance, or whether it is sufficient for obligations to be imparted through other sources, and internalized as part of the everyday operations of firms and activities of individuals (Hutter, 2001, pp. 6, 97; 2011, p. 43). Manufacturers' generally low (or lack of) awareness of their legal obligations for machinery safety, coupled with the finding that most firms were mediocre or poor performers for the substantive safety outcomes (see Chapter 3), may imply that knowledge of legal obligations is an important underpinning of compliance. For example, without a sound knowledge of the relevant legal obligations manufacturers had neither a conceptual framework for addressing machinery safety matters, nor a reliable understanding of what constitutes compliance.

The findings for detailed knowledge of legal obligations and exceptional performance lend some support to this conclusion, as four of the six firms that had a detailed knowledge of either the Australian or the European legal obligations for machinery safety were exceptional performers across all substantive safety outcomes. However, this coincidence between greater awareness of relevant legal obligations and better performance for safety outcomes is not, in itself, evidence of causation. Other factors may influence level of awareness or performance, or both of these. For example, an information seeking approach may support both a high level awareness of legal obligations, and sound decision making and action for substantive safety outcomes.

The finding, in Chapter 3, that although only a small proportion of study firms (9 per cent) met all substantive safety outcomes, a higher proportion performed well for at least one substantive outcome suggests that other factors were in play. How did 30 per cent of firms manage to comprehensively recognize the hazards for their machinery? What contributed to 47 per cent of firms using more effective safe place controls as the primary measures to eliminate or minimize risks? And how did 24 per cent of firms come to provide more substantial, good quality safety information for their machinery? Again a key part of the explanation lies in how manufacturers learned about machinery safety matters through practice as some of the sources of their learning, which went well beyond legal obligations, were linked to good performance for particular substantive outcomes, as discussed in Chapter 6.

A further consideration is that low awareness of the legal obligations for machinery safety did not mean that these obligations were without influence. As well as the three firms in which key individuals knew some detail about the relevant provisions in Australian OHS law, there were 11 other firms in which key individuals' awareness of these provisions was low but the perceived threat or authority of the law was the motivation for those firms to take some action to address machinery safety matters. In addition to the three firms in which key individuals knew some detail about the European regulatory regime for machinery safety, there were eight other firms in which key individuals' awareness of the European requirements was low but the law was a motivation for those firms to take some action to address safety matters in order for their machinery to be marketable in Europe. In addition, some firms were motivated to take action by a non-specific concern about litigation or legal liability. In some instances these legal imperatives were relayed or amplified by customers, distributors or other industry sources.

Legal obligations for machinery safety contributed to some firms' motivations for addressing safety matters, even though they did not necessarily compel firms to learn about specific obligations, accurately and comprehensively. The perceived threat or authority of the law could be the impetus to seek out, heed and apply information about machinery safety from other sources.

**Conclusion**

A major pre-occupation in legal standard setting is the design of regulatory provisions as their substance, precision and other properties may affect the capacity of regulatees to understand, interpret and apply their legal obligations (Black, 1997, pp. 22–3; Diver, 1983; Baldwin, 1995, pp. 7–11). These are important issues but it would be unhelpful to encourage the belief that changes to regulatory design alone would resolve the problem of low awareness of legal obligations demonstrated in this research.

In manufacturing firms, low awareness of legal obligations stemmed from disengagement from state regulation, with the requirements of the law often sitting outside the frames of reference and practice of those involved in machinery design and construction. Low levels of engagement with machinery safety legal obligations and, conversely, gleaning impressions (and sometimes misconceptions) about legal obligations through interactions with customers, distributors and other industry sources (or not at all), are part of the explanation of differences in business behaviour and performance explored in this book. They are part of understanding the constitution of knowledge and motivations, as key factors shaping business conduct.

The legal instruments dealing with the safe design and construction of machinery are, of course, only one half of state regulation. How regulators promote and explain, inspect and enforce the law may also impact on compliance. A crucial question then is whether regulators fostered the kind of self-regulatory action by machinery manufacturers that was necessary for them to comply with their ongoing legal obligations. The next chapter examines this issue, focusing on the compliance support, inspection and enforcement activities of OHS regulators, and manufacturers' experience of these. Chapter 6 then returns to the theme of learning through practice, explaining how firms, and key individuals in those firms, constructed knowledge about machinery safety matters through a much wider range of experiences grounded in everyday activities, and interpreted those experiences on the basis of their different personal histories and capacities. Chapter 7 illustrates how risk management practices took shape in different forms, in the context of each firm's operations and interactions with external actors, and Chapter 8 locates the legal obligations for machinery safety as one among a diverse set of motivations that shaped manufacturers' responses on safety matters.

# Chapter 5

# Encounters with Regulators

The occupational health and safety (OHS) statutes in Australia coupled obligations for the design and construction of machinery and other plant, with inspection and enforcement by the state, in a form of enforced self-regulation (Ayres and Braithwaite, 1992, pp. 101–6; Hutter, 2011, p. 12). Inspectors appointed under this legislation had (and continue to have) broad powers, to enter and inspect workplaces, investigate OHS matters, and receive information and assistance from those inspected (Johnstone 1997, ch. 7; 2004a, ch. 7; Johnstone, Bluff and Clayton, 2012, ch. 8). They could issue improvement notices requiring regulatees to remedy contraventions of the law, and prohibition notices to stop a work activity posing an immediate risk to health or safety. Failure to comply with such notices was a contravention of the OHS statutes. Inspectors could also initiate legal proceedings for non-compliance with a notice, and for suspected breaches of the designer and manufacturer obligations generally.

In Europe, the *Machinery Directive* provided for each member state to carry out market surveillance of machinery, in order to assess and enforce compliance with the directive (European Commission, 1998a,[1] Articles 1.2 and 7; see also Boy and Limou, 2003, pp. 83–99; Cordero and Muñoz Sanz, 2009; Cordero, et al., 2013; HSE, 2013a; 2013b; SWEA, 2002). Under the safeguard clause, a member state could initiate an administrative procedure to withdraw, prohibit or otherwise restrict marketing of machinery if it might endanger safety. The member state was required to inform the European Commission of the measures taken and the grounds for its decision, and must have sufficient technical evidence of non-conformance with the essential requirements or harmonized standards. The European Commission would then enter into consultation with the parties involved.

In principle, Australian-based manufacturers might encounter the compliance support, inspection and enforcement activities of safety regulators wherever they supplied their machinery, in overseas or local markets. In practice, the study firms had not encountered safety regulators in any countries other than Australia. This chapter therefore focuses on firms' experience of, and responses to, the activities of Australian OHS regulators. The analysis reveals manufacturers' minimal and spasmodic encounters with state regulatory activities, although closer encounters with regulators were more likely to capture firms' attention and prompt them to take action on machinery safety matters. The findings also point to weaknesses in regulatory policy, strategy and practice, as inspectors principally dealt with manufacturers reactively, treated compliance as a 'one-off' event rather than

---

1 Now European Commission 2006, Article 11.

ongoing, and infrequently applied the different types of enforcement mechanisms available to them. The influences of non-state actors in manufacturers' supply chains, networks and industry contacts, are also exposed as capable of heightening awareness of state regulatory activities, but at times miscommunicating information.

At one level the findings in this chapter constitute an account of regulatory enforcement in a particular context. However, machinery and equipment safety is a global concern and the issues raised here are food for reflection by regulators and policy makers working on market surveillance and enforcement issues in other national regimes, and in work or product safety generally. At another level this material offers readers with wider interests in regulation and socio-legal issues, insights into the relative impact of, and interplay between, state and non-state influences in shaping regulatees' knowledge, motivations and actions. For specialists and practitioners in human factors, engineering and design, the analysis offers a basis for questioning whether, or not, regulators in the regimes they deal with are reinforcing their efforts to enhance safety in the design and construction of machinery, and other items or materials.

The chapter considers, in turn, the different types of compliance support, inspection and enforcement mechanisms, and manufacturers' (or their customers' and distributors') encounters with these, before crystallizing the wider implications for state regulation. This is the need for regulators to carefully craft their interventions with firms as manufacturers, and the chapter finishes by introducing the key elements of an alternative approach; one which is networked, contextual, responsive and principled.

## Compliance Support

*Guidance and Advice*

To support compliance the OHS regulators published codes of practice and guidance materials, some of which included information relevant to machinery design and construction. They encouraged their staff to distribute or draw regulatees' attention to these information materials, which they made available through their websites and public offices. These materials provided information about the legal obligations for machinery safety, the risk management process, checklists to assist in identifying different types of machinery hazards and information about safeguards (see for example, Victorian WorkCover Authority, 1995; Workplace Services, 2000; 2002a; 2002b; 2002c; WHSQ, 2006; WorkCover NSW, 2007; 2008; WorkSafe Victoria, 2002). There were also alerts about fatalities and serious incidents, which warned of particular machinery risks and measures to minimize them.

Beyond providing codes and guidance materials, the regulators discouraged their staff from offering advice to any persons with obligations in OHS law, principally for resource reasons. As a manager with the Victorian regulator explained, "we're the only ones who can do the regulator role whilst there's lots of other players out there who can perform other roles … including that of consulting and advising duty holders" (Vic Regulator 7; see also Victorian WorkCover Authority, 2003b, ch. 1; Maxwell, 2004, pp. 260–2).

There was little evidence of manufacturers taking any action on machinery safety matters in response to regulator guidance materials or advice. Only six firms had sought advice from OHS regulators and, in line with the regulators' practice, had received guidance materials (9 per cent (6/66)). While these materials provided useful information they were not the stimulus for preventive action by firms. The findings about the low uptake and impact of guidance materials add to a growing number of studies which suggest that guidance material, whether in print form or on OHS regulators' websites, is not widely accessed and used by firms, particularly smaller ones (James, et al., 2004; Mayhew 1997a; 1997b; Mayhew, et al., 1997; Melrose, et al., 2006, p. iii; Wiseman and Gilbert, 2002; Wright, Marsden and Antonelli, 2004, pp. i, vii).

All but one of the six manufacturers that had obtained guidance materials were among a wider group of 28 firms in which key individuals complained about what they perceived to be a lack of, or poor quality, advice or information from OHS regulators (42 per cent (28/66)). This is a recurring theme in OHS regulation (Maxwell 2004, pp. 259, 262–5). Together, the findings for the low use of guidance materials and the perceived lack of advice from OHS regulators indicate that regulators were not key constituents of manufacturers' knowledge for addressing machinery safety matters.

*Compliance Programmes*

For a limited number of special purpose programmes the OHS regulators produced more tailor-made guidance materials. These were state-wide or national programmes conducted to address OHS problems which had wider ramifications for a range of businesses or industry sectors. Occasionally these programmes targeted producers or suppliers of particular types of machinery, and part of the intervention was the provision of machinery specific information.

An example was a programme conducted by the Victorian OHS regulator to address fatalities and serious injuries arising from forklifts (Lambert and Associates, 2003, pp. 3–21; Skinner and Stewart, 2006). The regulator had commissioned research, which revealed that risks previously attributed to poor driving by operators were actually caused by the unsafe design of forklifts. The research established that forklifts were highly susceptible to tipping over, could be overloaded to a level that impaired their stability, could be driven at speeds that exceeded their ability to stop without becoming unstable, and that pedals for braking, acceleration and reversing could be easily confused by operators. On the basis of this research, the regulator produced guidance material to clarify the state of knowledge about forklift hazards, and presenting a range of possible safe design and construction solutions to minimize the risks (WorkSafe Victoria, 2003a; 2003b). A senior manager explained:

> We weren't just saying, "here is a problem, what are the answers"? We were
> saying, "here is a problem because, and this is what you can do, now if you can't
> do that, how can you do it if you can't do these things?" And they were able to
> classify what would be a solution. (Vic Regulator 3)

The regulator used the guidance material in negotiations with forklift truck suppliers,
and with employers. These were aimed at securing industry support to incorporate
the identified solutions in forklifts supplied and purchased for use at work.

In a second example, the *Safer at the Source Intervention* in South Australia
targeted designers, manufacturers, importers and suppliers of agricultural
machinery (Workplace Services 1998a; 1998b). The regulator wrote to targeted
firms outlining the requirement for risk management, and advising them that an
inspector would visit or telephone to evaluate their risk management. An inspector
then visited a sample of firms to provide advice about risk assessment, and
disseminate guidance material about conducting risk assessments and producing
machinery safety information (Workplace Services, 1999; 2000). As such this
programme aimed to develop the capacity of upstream regulatees to implement
ongoing risk management, rather than one-off risk control outcomes.

The *Safer at the Source Intervention* was the only compliance programme
encountered by any of the study firms. Two firms were visited by an inspector as
part of this programme, and received guidance material as well as oral advice from
the inspector. In one of these firms, the owner/managing director learned about
machinery risk assessment from the inspector, and commenced a programme of
assessment using a procedure based on the regulator's guidance material. The
other firm did not take any action in response to the inspector's visit, believing
that they already had risk management in place for their machinery.

The impact of the regulators' programme-style interventions with machinery
manufacturers was limited as they only targeted firms producing or supplying
particular types of machinery. However, in targeting designers, manufacturers,
suppliers and/or end users, these interventions had the advantage of interacting
with a cross-section of parties in supply chains and markets for that machinery.
This strategy increased the potential to reinforce regulatory messages among
industry parties, and reduced the potential for resistance engendered by inconsistent
treatment of firms operating in the same markets. As discussed below, some study
firms resisted complying if they perceived that inspection and enforcement were
inconsistent and their competitors were not expected to comply to the same
standard of safety.

### Design Audits for Prescribed Machinery and Equipment

Firms that designed certain types of pressure equipment, cranes and hoists,
lifting equipment and some other types of machinery were required to submit
design documentation to the regulator in the state or territory where the item was

designed (see Chapter 2). For their part, the regulators audited some designs, but were not legally required to approve them. They reviewed the design calculations and drawings, and checked that the design applied relevant technical standards or engineering principles, and addressed safety critical aspects. They sometimes conducted an inspection of the machinery or equipment to verify that it was constructed in accordance with the design.

While the audit process was potentially an opportunity to raise awareness of safe design and construction issues with machinery designers, the regulators did not audit all submitted designs and, for those they did audit, they did not necessarily interact with the designer. Unsurprisingly, among the ten study firms that produced prescribed types of machinery or equipment, and submitted design documentation to the relevant OHS regulator, there was no evidence that any firm was influenced to pay greater attention to the safe design and construction of their machinery as a result of participating in this process. It is possible that the respective OHS regulators audited the design documentation from these firms, but none had interacted with an OHS regulator beyond submitting the documentation. In addition, some of the key individuals in these firms were not aware that the requirement to submit design documentation was established in OHS law, believing instead that it came from technical standards for boilers and pressure vessels, cranes or other items of machinery or equipment.

The process of submitting design documentation appeared to be too remote to capture manufacturers' attention and prompt them to take additional action on machinery safety matters, or to raise their awareness of the origin of the requirement in OHS law. In contrast some other, albeit limited, interventions by OHS regulators did prompt firms to take preventive action, at least in some circumstances, as set out in the next section.

## Inspection and Notices

### Direct Inspection

The OHS regulators operated within the traditional and dominant employer/ worker paradigm of OHS inspection and enforcement. That is, they prioritized the legal obligations of employers to protect their workforces, targeted employers with higher rates of injuries, and investigated incidents and complaints at employer workplaces (Victorian WorkCover Authority, 2000, pp. 14–15; 2002a, p. 2; 2003a, pp. 20–21; WRMC, 2002; 2006; see also Bluff et al., 2012). Inspection of firms as manufacturers, or their machinery, was therefore principally reactive and event-focused (Johnstone, 2003, chs. 3–5), occurring if the firm's machinery was involved in a work-related fatality, serious injury or incident, or there was a complaint about the machinery. Inspectors conducted little proactive surveillance of machinery, and principally at agricultural field days.

The machinery of 15 study firms had been inspected at their own, or customer or distributor sites, around Australia (23 per cent (15/66)). Eight firms' machinery had been inspected only once, and the other seven firms' machinery had been inspected on several occasions over a period of years. Some inspections were proactive (seven firms had experienced this). As the managing director of a firm that manufactured agricultural machinery stated:

> We will occasionally have an inspector go through a field day and just looking at machinery from a safety point of view and … that's the way to do it because all the machines are there and all the different manufacturers are there. And they've pulled us up a few times, a couple of times. (Managing director, Manufacturer 20)

This example illustrates the potential for OHS regulators to increase their proactive surveillance of machinery at field days, trade shows, distribution facilities and other similar situations where producers and suppliers display new machinery. However, the study firms were more likely to experience inspection reactively in response to an incident involving their machinery (eight firms), or an end user complaint about the machinery (two firms).

In their dealings with machinery manufacturers inspectors favoured a cooperative approach (also called a compliance or accommodative approach; see Black, 2001b; Hutter, 1997, pp. 15–16). Such an approach is typically characterized as involving advice and persuasion (Hutter, 2001, pp. 15–16; Johnstone, 2004a, pp. 150–151) but manufacturers seldom encountered inspectors who were more facilitative (May and Wood, 2003), and passed on information about machinery safeguarding or other matters. Rather, inspectors gave directions and sought to persuade firms to take preventive action. The efficacy of persuasion in the absence of advice is called into question by empirical research, which suggests that unless regulatees already know how to comply, the regulator will need to play a more educative role (Fairman and Yapp, 2005a; May and Wood, 2003).

Manufacturers' accounts of inspection indicated both the preference for persuasion and a reluctance to issue notices. The manager of a firm that manufactured a grain cleaner explained the approach taken by an inspector in asking the firm to guard the machine, after a farmer's fingers became trapped in an unguarded belt:

> There was no "you must change this". It was "we would like to see you change it" and we've conformed but there was no actual regulatory enforcement. There wasn't an improvement notice given to us or anything like that. It was just a suggestion was made and we worked together … they said, "oh, this is what we would think you should do, this needs to be guarded and that needs to be guarded", but it was up to us to develop the guards to suit and then when we believed we had it satisfactory they came back and had a look and they were happy. They said, "you could do this better or you could do that better" and we explained, "well, we can't do that because that gets in the way", and they said, "oh, that's fine then". (Manager, Manufacturer 33)

Like most of the manufacturers whose machinery had been inspected, this firm was not issued with a notice in relation to the grain cleaner. Inspectors rarely shifted from cooperation to insistence (Hutter 1997, p. 16), by issuing a notice, in order to formalize their communications and increase the pressure for firms to take preventive action, if they had failed to do so. Overall, no firms had received or were aware of prohibition notices for their machinery, and only two had been issued with improvement notices. A third was aware of an improvement notice issued to a distributor concerning the firm's machinery.

All three notices related to risk assessment. For example, in one instance an inspector responded to a complaint by workers at a construction site about the manufacturer's crane. As the firm's production manager explained:

> ... the WorkCover[2] guy came in and he put certain requirements on them [management], relating to risk assessment being done on cranes.

> ... we were issued with a WorkCover improvement notice to do just that.
> (Production manager, Manufacturer 47)

In this case the inspector investigated the complaint and issued an improvement notice to the crane manufacturer to conduct comprehensive risk assessments for each of its cranes. The company previously had a single generic assessment for all of the firm's cranes.

As well as requiring manufacturers to conduct or improve a risk assessment for the machinery, the type of preventive action sought by inspectors focused on addressing particular design faults, incorporating or improving interlock systems, guarding or other risk control measures, or providing or amending safety information. As such, inspectors treated compliance by these firms as a single (one-off) event, rather than fostering the commitment, capacity and arrangements to ensure ongoing attention to safety in machinery design and construction (Johnstone and Jones, 2006, pp. 485–6; Parker, 2002, pp. ix–x, 43–61).

Of the 15 manufacturers whose machinery had been inspected, ten firms, including the three that had received or were aware of improvement notices for their machinery, had taken some preventive action that was attributable wholly or in part to the inspection (67 per cent (10/15)). Although inspection was rare, it had the potential to capture management's attention and prompt some action to incorporate guarding or other risk control measures (11 examples), conduct a risk assessment (four examples), or improve machinery safety information.

---

2  WorkCover was the name of the combined workers' compensation and OHS agency, of which the OHS regulator was part.

*Awareness of Inspection through Intermediaries*

As well as direct experience of inspection, some manufacturers had encountered or were aware of the potential for inspection and enforcement more generally, through their interactions with customers or distributors (15 per cent (10/66)). As one manager stated, "I have seen them go in and stop machines that are new in there because of inadequate guarding" (Manager, Manufacturer 23). Messages about inspection and enforcement relayed through customers or distributors also encouraged some manufacturers to give priority to safety matters, because these parties might not accept machinery if it was unsafe and could be subject to enforcement in end use. As the technical services manager of a firm manufacturing construction machinery stated:

> The people that are running the businesses in the industry that we're in are more perceptive now of what they have to buy, and they look at it from all angles because if they go on a building site, a lot of them are employers, and they know that if something happens then WorkCover or whatever is going to go totally through the roof. So they're putting us under pressure too. (Technical services manager, Manufacturer 12)

This example illustrates the way that manufacturers' supply chain interactions could contribute to them taking action on machinery safety matters. Among the ten firms aware of inspection and enforcement at customer or distributor sites, seven were motivated to take some preventive action as a result of this intervention (70 per cent (7/10)).

*The Potential for Resistance*

Whether manufacturers experienced inspection and enforcement directly, or were aware of the possibility of inspectorate involvement through their customers or distributors, there were limits to the action that they were willing to take. This was especially the case if they perceived that particular measures would impede or conflict with their economic goals relating to the marketability of their machinery and firm profitability. As explained further in Chapter 8, manufacturers were motivated by a series of economic concerns, which shaped their responses on machinery safety matters. These concerns included optimizing the functionality of machinery, containing costs and avoiding a competitive disadvantage. Such motivations provided manufacturers with justifications for limiting or not taking action on machinery safety matters, even in the context of directions from OHS inspectors.

A manufacturer of grain auger conveyors provided an example of resistance to compliance with an inspector's direction, because of over-riding concerns about machinery functionality. The firm's managing director stated that:

An auger-type conveyor is no good unless you can get the product into it. In order to do that at some stage you have to have openings large enough for the average product to go into the moving part. They [inspectors] came along and said we need a mesh on here that won't let anybody get access to it. Well fine. Then just outlaw them as a form of conveyor. Why? Well because if you can't get access to that bit then the machine can't work. It simply can't do what it was designed to do ... the mesh on the hopper isn't small enough to stop the people from getting a hand through it. It stops you falling into it or anything else but it has to be an opening that will let the grain run through it. (Managing director, Manufacturer 16)

This managing director perceived that complying with the inspector's direction to fit finer mesh or other guarding on the grain auger would impede its functionality and, in turn, its marketability. The economic concern relating to functionality constituted a justification not to implement risk control measures.

A second example, involving a particular type of press, illustrates the extent to which manufacturers' preferenced machinery marketability over the need to effectively control known and serious risks. In the context of inconsistent inspection and enforcement by OHS regulators with other manufacturers of the same type of machinery, the press manufacturer resisted implementing effective controls on the grounds that the firm would be placed at a competitive disadvantage. Following a coronial investigation into a fatality in the state of New South Wales involving the press, the manufacturer had negotiated with the OHS regulator in that state to gain acceptance for particular risk control measures for this type of press. The solution negotiated with the regulator completely guarded the press to prevent access to a hydraulic ram. The design was inherently safe as the ram would not operate if a guard was open, eliminating the possibility of crushing injuries. The manufacturer believed that as a result of these negotiations with the regulator, the agreed solution would be enforced with all manufacturers and suppliers of this type of machinery, in all Australian states.

The press manufacturer developed a new model which incorporated the inherently safe design solution. This was done at some considerable cost to the firm (about $A500,000), to test, tool up, build and supply the new model, and due to a loss of market share when competitors continued to supply their less safe models of the press and OHS regulators did not require them to do otherwise. The manufacturer then opted to supply a lower cost model, equivalent to its competitors, which would prevent a fatality but would not prevent serious injury, and marketed its inherently safe model as an option for customers who were prepared to pay the additional cost for safety. The owner and managing director of the firm explained:

... we worked very closely with the NSW WorkCover Authority and they came down and inspected the machine in as much as they can ever, or are ever prepared to pass things, but they sort of gave their nod of approval but never on paper. They were very happy with it and we thought great that level of guarding will now become the benchmark if you like or the standard. Not so much the Australian Standard but the standard by which WorkCover would apply rules.

... Despite us forwarding all of the details to the national bodies and saying well "hey, we've done this let's just have a level playing field". We saw a copy of the letter ... in relation to our competitor in Western Australia to say an [alternative mechanism] was acceptable ... which I would call a secondary safety device, that you activate when you need to, it's not a primary one that keeps you out ... If you're caught, as your arm's pulled into the machine, as you come down, in the worst case your arms would press this bar and stop the machine before they sort of got broken off ... it's an effective tool to save death if you like, but it doesn't prevent someone from getting in, in the first place.

... and just through commercial survival we were forced back to that level ... we have to this day still offered a fully guarded machine ... but we also offer another model which is the alternative version and that's by far the highest volume that we produce because it's the cheaper one of the two. (Owner/managing director, Manufacturer 68)

The press example indicates the perils of regulators seeking to secure compliance by one firm at a time, and either not interacting with others producing or supplying the same type of machinery, or treating like problems differently. Failing to pay attention to fundamental principles of regulatory implementation (Yeung, 2004, pp. 36–43), by inconsistent treatment of firms competing in the same markets, is a risky approach for regulators, as perceived unfair treatment is more likely to meet with resistance and to impair the regulator's legitimacy (Murphy, 2005; Tyler, 1997, 2006; and see also Bardach and Kagan, 2002). The potential for resistance engendered by manufacturers' over-riding concerns about the marketability of their machinery, competition and business profitability, highlights the need for OHS regulators to carefully craft their inspection and enforcement strategy to take account of economic motivations. Such a strategy is outlined later in this chapter.

Beyond inspection, with or without notices, inspectors could be more coercive by using a deterrence or sanctioning approach (Black 2001b; Hutter, 1997, pp. 15–16). This would involve formally investigating non-compliance and proceeding to prosecution, as discussed below.

## Criminal Enforcement

The OHS regulators could bring legal proceedings for contraventions of any provisions of their OHS statutes or regulations. In the event that a prosecution was successful, the court could impose a penalty with the principal penalty being a fine[3] (Johnstone, 1997, pp. 401–5; Johnstone 2004a, pp. 445–50).

*Specific Deterrence*

None of the study firms had been prosecuted in relation to the design and construction of their machinery, but one firm's machinery had been the subject of investigation by an OHS regulator after a worker died while using the machinery. At the time of data collection, the firm was aware that it might be prosecuted. The fatality, the ensuing investigation and an improvement notice issued by the regulator about the risk assessment for the machinery had contributed to the firm reviewing the assessment, changing guarding and revising the safety information provided. Being investigated for prosecution reinforced the owner/managing director's perception of inconsistent enforcement by OHS regulators as the firm's competitors had not been subject to enforcement. He explained the firm's current response to machinery design and construction matters:

> ... while we're improving the safety of the machine we will enhance its functionality or usability ... we will no longer accept a regression of the usability of the machine ... Wherever possible we don't just put the guard on and say that's an impost ... we try and enhance the machine by doing that. And that's difficult but it's a bit of a change in philosophy of late. (Owner/managing director, Manufacturer 68)

This manufacturer was responding to regulatory pressure, experienced over a period of years, on the manufacturer's terms. The firm would not implement safety measures that impeded machinery functionality or rendered the firm less competitive in the market for its machinery. They had developed inherently safe designs for three different types of machinery in order to reconcile functionality requirements and enforcement pressures for greater safety.

Another manufacturer had taken more ongoing steps to manage machinery safety matters as a result of prosecution for a breach of the firm's duty of care as an employer, coupled with economic concerns about workers' compensation costs. The prosecution was instrumental in influencing the firm to systematically manage health and safety matters in general, including attention to safety in machinery design and construction. The increased commitment to safety arose from the firm's desire to offset the stigma of prosecution, which had the potential to adversely impact the firm's plans to gain financially beneficial status as a self-insurer for workers' compensation (see also WorkCover Corporation, 2001a, p. 16).

These two instances of manufacturers with direct experience of prosecution or investigation for prosecution were exceptional. It is not possible to draw wider

---

3 A wider range of orders are now available, in addition to financial penalties (see Chapter 9, ''Implications for State Regulation of Machinery Safety').

conclusions from them about the specific deterrent effects of these actions by OHS regulators. The examples are, however, consistent with regulatory theory and empirical research which suggest that inspection or investigation accompanied by sanctions can focus managerial attention and lead to improvements in an organization's performance (see for example Baggs, Silverstein and Foley, 2003; Fan, Foley and Siverstein, 2003; Fan, et al., 2006; Gray and Mendeloff, 2002; Gray and Scholz, 1990, 1991, 1993; Gunningham, Thornton and Kagan, 2005; Ko, Mendeloff and Gray, 2010; Mendeloff and Gray, 2005; Shah, Silverstein and Foley, 2003; Weil, 1996, 2001). The two examples also suggest that deterrent effects may not be realized unless firms align or frame enforcement by regulators as consistent with their other business goals.

*General Deterrence*

At the time of data collection for the research, only some of the OHS regulators had pursued legal proceedings with businesses as machinery designers, manufacturers or suppliers, and only infrequently. The Victorian regulator had successfully[4] prosecuted a manufacturer/supplier of a can press in the case of *Hydrapac*,[5] and there have been several subsequent prosecutions of machinery or equipment designers and/or manufacturers. In addition to *Hydrapac*, the other prosecutions are the cases of *Outdoor Initiatives*,[6] *Tornado Pumps and Sprayers*[7] and *Jalor Pty Ltd*,[8] in which the respective defendants were the designer/manufacturer of a flying fox,[9] the manufacturer of an agricultural spray pump and the designer/ manufacturer of wood working machinery. None of these cases established important precedents for the safe design and construction of machinery. Each followed a fatality or serious injury involving a designer and/or manufacturer of machinery or equipment, and dealt with the circumstances leading to the incident, basic safeguarding issues or the adequacy of safety information, and did not address the firms' ongoing management of machinery safety matters. In each case sentencing was unremarkable, with fines respectively of $A7,500, $A20,000 and $A40,000 (the latter for each of two counts).

---

4   For discussion of the earlier, unsuccessful prosecution of Chem–Mak, and its dampening effect on prosecution of designers, manufacturers and suppliers in Victoria, see Bluff, et al. (2012).

5   *Victorian WorkCover Authority v Hydrapac Pty Ltd* (unreported, Magistrates Court (Vic), McDonald M, 25 October 2001).

6   *Victorian WorkCover Authority v Erik Westrup trading as Outdoor Initiatives* (unreported, Magistrates Court (Vic), Crisp M, 20 February 2006).

7   *Victorian WorkCover Authority v Tornado Pumps and Sprayers Pty Ltd* (unreported, Magistrates Court (Vic), Wright M, 5 May 2006).

8   *Victorian WorkCover Authority v Jalor Pty Ltd* (unreported, County Court of Victoria, Coish J, 25 January 2010).

9   A flying fox is an overhead cable and apparatus for transporting materials over difficult terrain.

There was also a small number of prosecutions of suppliers or importers of machinery and equipment which, like the designer/manufacturer cases did not involve design or construction issues.[10] They were also event-focused, dealt with basic safeguarding or information provision matters, and were not significant for the ongoing self-regulation of safety in machinery design and construction.

Similarly, in Queensland there were few prosecutions of designers or manufacturers of machinery, and with modest penalties. Two cases resulted in fines less than $A25,000 without conviction, and a third was adjourned without conviction together with a good behaviour bond.[11]

A more influential case was that of *Arbor Products*,[12] in which the New South Wales OHS regulator prosecuted the manufacturer/supplier of a wood-chipping machine for failing to ensure the machine was safe and without risks to health when properly used. Arbor Products supplied the machine to a shire council and, at the time of supply, had provided an operating manual and training to some of the council's employees. An employee sustained traumatic amputation to both arms when he was drawn into the chute of the machine and his arms came into contact with the cutting blades. The injured employee had not received training and did not have access to the manual.

The key finding of the Industrial Relations Commission (IRC) of New South Wales in full session (on appeal) was that the statutory duty required that machinery supplied was safe, in the sense that its safety was ensured. The wood-

---

10   The cases are *Inspector Arnott v Wreckair Pty Ltd* (unreported, Dandenong Magistrates Court (Vic), 30 October 1991); *Victorian WorkCover Authority v Anton's Mouldings Pty Ltd* (unreported, Dandenong Magistrates Court (Vic), Harding M, 5 June 2001); *Victorian WorkCover Authority v Melbourne Cranes Imports Pty Ltd* (unreported, Melbourne Magistrates Court (Vic), Crowe M, 12 September 2001); *Victorian WorkCover Authority v Marco Packing Machine Supplies Pty Ltd* (unreported, Ballarat Magistrates Court (Vic), Bolster M, 25 November 2002); *Victorian WorkCover Authority v East End Hire Pty Ltd* (unreported, Melbourne Magistrates Court (Vic), Hannan M, 24 June 2003); *Victorian WorkCover Authority v Mastaquip Pty Ltd* (unreported, Melbourne Magistrates Court (Vic), Hannan M, 16 April 2003). For later supplier cases, after the period of data generation for this research, see *Victorian WorkCover Authority v Press Shop Automation Pty Ltd* (unreported, Melbourne Magistrates Court (Vic), Walter M, 10 March 2005); *Victorian WorkCover Authority v Nick And333niko (T/A High Five Jumping Castles)* (unreported, Melbourne Magistrates Court (Vic), Popovic M, 14 December 2006); *Victorian WorkCover Authority v Phil Panzera (T/A High Five Jumping Castles)* (unreported, Melbourne Magistrates Court (Vic), Popovic M, 14 December 2006); *Victorian WorkCover Authority v Extec Sales and Distribution Australian Pty Ltd* (unreported, County Court (Vic), Coish J, 19 June 2009).

11   The cases are *Simville Pty Ltd* (Unreported, 28 September 2005); *Rodney Charles Sammon* (Unreported, 5 December 2005); and *Air Design Pty Ltd* (Unreported, 16 December 2010). The few details available for these cases were provided by Graham Lee at Workplace Health and Safety Queensland.

12   *WorkCover Authority of New South Wales (Inspector Mulder) v Arbor Products International (Australia) Pty Ltd* (2001) 105 IR 81.

chipping machine was inherently unsafe as the chute for feeding material was too short and permitted contact with rotating blades. The IRC also found that the manufacturer/supplier's legal liability was limited only where machinery was safe but became unsafe because of misuse, but not where the defendant had failed to make the machine safe and had only provided an instruction manual, advice or training in the proper use of the machine. Arbor Products was fined $A30,000. A number of other New South Wales cases have subsequently confirmed the principle established in *Arbor Products*; that machinery and other plant must be designed and constructed to be safe, and that provision of information or training are not a substitute for this.[13]

The principle of ensuring that machinery is designed and constructed to be safe was also endorsed in the Western Australian case of *Viticulture Technologies*.[14] This prosecution of the supplier of a grape harvesting machine followed an incident in which a worker sustained serious spinal injuries when he fell from the top of the machine, while he was attempting to clear a blockage. The Stipendiary Magistrate in the case concluded that going to the top of the machine to clear blockages was foreseeable and within the scope of proper use, and that it was practicable for the harvester to be fitted with safety platforms so as not to expose workers to the risk of falling. He also indicated that while not binding in Western Australia, the *Arbor Products* case set the appropriate boundaries, and fined Viticulture Technologies $A20,000.

A key reason for the regulators' minimal prosecution of designers and manufacturers (and suppliers) was the practical difficulty of investigating and gathering evidence across state/territory borders, in instances where unsafe machinery was designed and/or constructed in one jurisdiction, and then supplied for end use in another. In principle, however, a prosecution could succeed if a territorial nexus was established, either through a provision in a relevant statute that enabled this, or by establishing a real connection between an offence and the state. This was well established in the New South Wales prosecution of a Victorian

---

13    For New South Wales cases confirming the approach taken in *Arbor Products* see *National Hire Pty Ltd v Howard* [2003] NSWIRComm 144 (9 May 2003); *Inspector Wilkie v Batequip Pty Ltd (formerly Bateman Equipment Pty Limited) T/as Ditch Witch Australia* [2003] NSWIRComm 111 (14 April 2003); *Inspector Batty v Vehicle Inspection Systems Pty Ltd* [2004] NSWIRComm 19 (27 February 2004); *Inspector Wilkie v Kennards Hire Pty Ltd* [2004] NSWIRComm 167 (10 June 2004); *Inspector Ruth Buggy v Kentan Pty Ltd* [2005] NSWIRComm 152 (12 May 2005); *Inspector Ruth Buggy v Lyco Industries Pty Ltd* [2005] NSWIRComm 423 (24 November 2005); *Lyco Industries Pty Ltd v Inspector Ruth Buggy* (WorkCover Authority of New South Wales) [2006] NSWIRComm 396 (13 December 2006). For a rare, and more recent, prosecution of a designer of plant see *Inspector Ching v Simpson Design Associates Pty Ltd* [2009] NSWIRComm 213 (15 December 2009).

14    *Shepherd v Viticulture Technologies (Aust) Pty Ltd* (unreported, Court of Petty Sessions, Albany (WA), Malone SM, charge no 1941/01, 15 May 2003).

manufacturer/supplier in the case of *Lyco Industries*[15] (see also Bluff, et al., 2012; Johnstone, Bluff and Clayton, et al., 2012, ch. 8).

The *Arbor Products* and *Viticulture Technologies* cases were, however, the only ones that some study firms were aware of and there was little evidence of firms taking action on machinery safety matters as a result of these cases. The key individuals in seven firms had heard of the *Arbor Products* case and one knew of the *Viticulture Technologies* case. In a ninth firm, the general manager was aware of both these cases. He had an accurate understanding of the key findings in the two cases, and his was the only firm to have taken constructive preventive action consistent with the findings in these cases.

The positive response in this firm came about as they had been asked to redesign a wood-chipping machine, which had similar design problems to the machine supplied by Arbor Products. As well as learning about the two cases through the industry association representing manufacturers and distributors of agricultural machinery (the Tractor and Machinery Association (TMA)), this general manager had obtained and read the *Arbor Products* case transcript in order to inform himself about the issues in redesigning the wood chipper. He explained:

> ... we were approached about whether we'd like to take these wood chippers on. We said well to do that we'll look at what needs to be done, so I contacted TMA and they've given me all the information. It's quite a lengthy document [the transcript] ... that one's the actual appeal in regards to it but it covers everything that happened in the court case. You can see I've gone through it [referring to highlighting and annotations].

> ... we are now in the process of designing it based on recommendations that are out there, especially in regards to the Arbor case, making that wood chipper say fool proof I suppose.

> ... operator manuals even though it's a good idea, you can say ... "do not do this", "do not do that", you can have warning stickers but ultimately you have to create or design a machine that can prevent them from doing that. (General manager, Manufacturer 48)

From the *Arbor Products* and *Viticulture Technologies* cases, this general manager understood that his firm must make the wood-chipping machine inherently safe and could not rely on warning end users of risks. He was redesigning the machine, including extending its chute so that an operator could not physically access rotating blades by reaching into or being drawn into the chute.

In the other eight firms in which the key individuals were aware of either the *Arbor Products* or *Viticulture Technologies* cases, industry associations or other

---

15　*Lyco Industries Pty Ltd v Inspector Ruth Buggy* (WorkCover Authority of New South Wales) [2006] NSWIRComm 396 (13 December 2006).

industry intermediaries were the source of information, and these individuals knew nothing about the findings in the cases or had misconceptions about them. For example, the managing directors of two firms learned of the *Arbor Products* case through information distributed by their industry association (the TMA). From this they were aware that the TMA considered the case to be a threat to the commercial viability of machinery manufacturers and suppliers, but they did not know precisely what action was indicated by the findings in the case. The managing director of one firm had registered that there was a need to 'idiot proof' machinery. He stated:

> Now what I've been reading from the Tractor Machinery Association they're just saying every manufacturer is just going to close their door. I don't know what law they changed but it really opened up the floodgates they reckon … we actually sat out there and we were looking at something, whatever it is and I said to my other partner here, "well what can go wrong", and we're running through it and I'll say, "oh you could get your fingers caught in there" or something and he'll go, "if you're a bloody idiot" and I said, "well you've really got to allow for bloody idiots". (Managing director, Manufacturer 3)

The managing director of the second firm was aware of the need to take action, and was willing to do so, but was taking action out of step with that indicated by the *Arbor Products* case. Rather than producing machinery to be inherently safe and providing safety information to supplement this, he mistakenly believed that the required action was to produce an operator manual and make sure his customers were aware of this by going through it with them. This managing director explained his perspective:

> I get newsletters from the Australian Tractor and Machinery Association of people that have bought a particular wood-chipping machine … . They were supplying it with an instruction manual, safety instruction, all that sort of thing. They sold it through a distributor and the distributor sold it to a customer who used it, who had an operator that fed his arm into the machine and got it chopped off. Consequently the manufacturer was found to be at fault because they never instructed the distributor on how to use the machine, and who'd never then instructed the end user on how to use the machine.

> … we generally try to deliver the machines ourselves and also instruct, assemble the machine with them, show them how to maintain it … all of that sort of thing and then go through the instruction book with them because of this Tractor and Machinery Association thing – that case … We produced a manual as a result of that information, we never did it before. (Managing director, Manufacturer 32)

These two managing directors' understandings of the *Arbor Products* case were distorted by the industry association's campaign at the time. The association argued that much agricultural machinery was dangerous if misused, and that employers, owners and operators of machinery should be held responsible for misuse, rather than designers and manufacturers being required to make machinery safer (Tractor and Machinery Association, 2001a; 2001b; 2001c; 2002). The TMA information stated that Arbor Products' wood chipper was fitted with industry accepted guarding, and did not highlight the need for the chipper to be inherently safe by making the chute longer so as to prevent access to rotating blades. The Association's stance on the *Arbor Products* case helped make its members aware of this prosecution, but did not assist them to accurately understand the key findings in the case.

A third managing director knew of the *Arbor Products* case from a different industry association. He also believed that machinery manufacturers should not be penalized for what he perceived to be operator misuse of machinery. He was poorly informed about the case and responded with defiance stating:

> The one that's I guess affected most machinery dealers of our style and size in Australia recently, is the people who built the ... machine for mulching tree branches. Now the machine was built with goodness knows how many guards to stop the bloke getting in and it had instructions, it had instruction books, it had big stickers on it and everything else. But an operator in Sydney took off, disabled and dismantled enough of the safety guards, ignored everything else and finally succeeded in hurting himself with the machine and quite rightly, as I would have done, the manufacturer objected to being blamed for that. And yet when it went through the system the ultimate thing was, "you created it, inherently it could hurt someone, therefore you're to blame" ... that means that anybody who creates any sort of machinery is inherently to blame. You cannot have that sort of thing hanging over manufacturers and expect innovation and investment. (Managing director, Manufacturer 16)

These examples of awareness of and responses to the *Arbor Products* case suggest that industry sources were not a reliable way for manufacturers to learn about the findings of cases. Some safety and socio-legal scholars propose that regulators can harness intermediaries such as industry associations, customers and parties in supply chains as ways to communicate regulatory messages to firms, through sources which firms trust and interact with regularly (Cowley, 2006; Gunningham and Sinclair, 2002, pp. 17–18; Hopkins and Hogan, 1998; Lamm and Walters, 2004, pp. 103–5; and see Gunningham, 2010, pp. 131–5; HSE, 2002; Wright, et al., 2005). A contrary view is that the potential for health and safety information to be communicated through intermediaries may be limited (James, et al., 2004; see also Hutter and Jones, 2007; Walters and James, 2009). The present research suggests caution in relying on intermediaries, as information may be miscommunicated or distorted through these channels, and the outcomes may be counterproductive for safety.

Empirical research on general deterrence across different regulatory regimes suggests that firms' awareness of enforcement of others in an industry can remind them to check their compliance, but that general deterrence does not occur if firms are not aware of enforcement because it is too infrequent, especially in the context of competing business communications (Fairman and Yapp, 2005a; Gray and Scholz, 1990; Gunningham, Thornton and Kagan, 2005; Jamieson, et al., 2010; Thornton, Gunningham and Kagan, 2005). Also, general deterrence does not occur if firms hear about enforcement second or third hand, through word of mouth, and do not have sufficiently reliable details to know what action they should take. Further, general deterrence may be less effective if differences in the nature of business operations make comparisons difficult; that is, firms do not identify others experiencing enforcement as 'people like us'.

In the present research, general deterrence was not effective for most firms because they were simply not aware of prosecutions against other manufacturers (or suppliers) (86 per cent (57/66)). The small number of firms that were aware of the *Arbor Products* or *Viticulture Technologies* cases (14 per cent (9/66)) had no difficulty identifying with these firms as companies like their own, and some were prompted to check their own performance in some respect. Only one firm had information through a primary source because the general manager had obtained and read the *Arbor Products* case transcript. The other firms were aware of one or other of the prosecutions through intermediaries, and received second or third-hand information about cases. They did not have sufficiently reliable details of cases to know what action they should take and the accounts they received reinforced their underlying attitudes that unsafe end users should take more responsibility for safety, rather than the manufacturer. This attitude was prevalent among study firms and mediated their responses on machinery safety matters, as examined further in Chapter 8.

With so few Australian prosecutions of Australian firms as designers or manufacturers there is a case to be made for OHS regulators to make greater strategic use of prosecutions. This research suggests that more prosecution, in and of itself, is unlikely to stimulate effective action to address machinery safety. For OHS regulators there is a challenge to determine ways to ensure that the key findings of prosecutions are effectively communicated to regulatees, across jurisdictions, and to clearly and succinctly set out the implications of particular cases. Regulators will also need to take account of third party intermediaries' communications about cases and regulatees' other motivations, which may influence whether they hear and absorb key messages about prosecutions.

## Issues in Compliance Support, Inspection and Enforcement

With broad statutory powers for inspectors to enter workplaces, investigate safety matters, receive information from regulatees, issue notices and initiate legal proceedings, there was the promise of OHS regulators being a potent influence on manufacturer's performance for machinery safety. The preceding discussion has shown, however, that there are many reasons why state regulation may have considerably less impact than anticipated. Key issues are the minimal, spasmodic and typically reactive engagement with firms as designers and manufacturers (or suppliers), the reliance on guidance materials and persuasion, and settling for one-off responses to inspection rather than action to manage safety in machinery design and construction on an ongoing basis.

Notwithstanding these weaknesses, the indications are that closer encounters between regulators and manufacturers, through direct dialogue, inspection and prosecution, are more likely to capture firms' attention and prompt them to take some form of preventive action. On the other hand, significant challenges are manufacturers' low uptake and actioning of the regulators' own guidance materials, the potential for miscommunication or misinterpretation of regulatory messages relayed through non-state intermediaries, and firms' economic motivations which may engender resistance to compliance if regulatory interventions unevenly disadvantage a firm compared with others in its markets.

All of this points to the need for a rethink of regulatory policy, strategy, planning and priority setting, and resource allocation, and questioning of well-worn ways of doing things. Firmer foundations may lie in interventions that are networked, contextual, responsive and principled.

Applying a *networked* approach, regulators would target a cross-section of manufacturers and other relevant regulatees, such as suppliers, importers and end users of the same type of machinery, in order to reinforce the need to comply across markets, promote a level playing field and overcome firms' resistance to compliance due to competitive pressures. A *contextual* element would encourage inspectors to tailor enforcement to regulatees' capacities, motivations and circumstances. It would seek to harness, and when necessary disarm, the contribution of industry and other non-state actors to manufacturers' knowledge and motivations for addressing machinery safety matters, including countering the potential for misunderstandings and resistance to be fostered by these parties.

A *responsive* approach would see inspectors nurturing willingness and capacity to comply through advice and persuasion, signalling the potential to apply sanctions if this cooperative approach fails, and progressively escalating through a hierarchy of mechanisms from notices to prosecution, if non-compliance persists[16] (Ayres and Braithwaite, 1992, pp. 35–41; Braithwaite J, 2011; Johnstone 2004b, pp. 155–60). The *principled* element would aim to ensure that regulators' interactions with manufacturers are consistent, transparent and procedurally fair (Yeung, 2004, pp. 36–43), again to reduce the potential for resistance to compliance associated with uneven inspection and enforcement of a firm and its competitors.

---

16   Under risk-based variants of this strategy inspectors respond based on a firm's self-regulation of safety matters, its past performance, and the degree of risk resulting from identified contraventions (Black and Baldwin, 2010; Gunningham and Johnstone, 1999, pp. 123–9; Johnstone, 2004b, pp. 159–60).

Regulatees are more likely to regard an authority as legitimate and to cooperate with that authority, if the latter exercises its powers in an impartial way, and is perceived to be fair (Murphy, 2005; Tyler 1997; 2006). For genuine consistency and transparency OHS regulators would also need to coordinate their activities across borders (see for example HWSA, 2007).

## Conclusion

The findings in this chapter are significant in three ways. First, they contribute to understanding how machinery manufacturers constructed knowledge about machinery safety matters, albeit in the negative. They indicate the low uptake and application of the OHS regulators' principal mechanism for supporting compliance, their guidance materials, and manufacturers' low opinion of OHS regulators as a source of advice and information. This adds to the findings about low awareness of legal obligations for machinery safety among the key individuals in manufacturing firms (Chapter 4). Taken together, these findings demonstrate that state regulatory sources were not key constituents of knowledge about safety matters in firms.

Second, the findings are significant in locating action by OHS regulators in the context of manufacturers' other motivations for taking, or not taking, action on machinery safety matters. The chapter has shown that inspection and enforcement by OHS regulators were among the motivations for some firms to take preventive action, but co-existed with economic motivations which reinforced or conflicted with regulatory pressures. Chapter 8 explores this issue further, showing that manufacturers' responses on machinery safety matters were shaped by a series of motivational factors, some of which supported and some of which impeded firms taking preventive action, or adversely impacted the quality of that action.

Third, the findings in this chapter have significance for public policy. Legislation that is not enforced effectively rarely fulfils its social objectives (Gunningham, 2010, p. 120), and this research points to the need for OHS regulators to take a fundamentally different approach in their interventions with machinery manufacturers. They will need to tackle the competing motivations of firms and gaps in their knowledge, and place greater emphasis on firms' commitment, capacity and arrangements to comply with OHS law on an ongoing basis. As introduced here, and clarified further in Chapter 9, regulators may best support and enforce safe design and construction of machinery through an approach which is networked, contextual, responsive and principled.

This chapter concludes consideration of the nature and influence of state regulation, in the form of legal instruments and the activities of OHS regulators, on whether and how manufacturers addressed machinery safety matters. The chapter has confirmed that regulation by the state may play a positive role in shaping regulatees' knowledge and motivations but, for machinery manufacturers, this was relatively minor. Moving beyond state regulation, a second conceptual theme of interest in understanding compliance is organizational capacities, which may affect business decision making and implementation. The next chapter takes up this conceptual theme, examining how and what manufacturers learned about machinery safety matters, and the implications of this for their performance and compliance.

# Chapter 6
# Practice, Learning and Performance

The preceding chapters have established manufacturers' disengagement from, or minimal exposure to, state regulation in the form of the Australian and European legal obligations for machinery safety, and the activities of occupational health and safety (OHS) regulators. This finding might be taken as an explanation for the mediocre or poor performance of some firms for hazard recognition, risk control and provision of safety information. The comparatively small role of state regulation is, however, only a partial explanation for the mixed performance of firms.

This chapter begins the process of examining other conceptual themes of interest in understanding and explaining manufacturer performance. It looks at the development of capacity in firms, which research into safety management and corporate self-regulation more generally has shown is crucial to achieving social and business goals (Gallagher, 1997, s.5.6; Hale and Hovden, 1998; Hutter, 2001, pp. 301–2; Nytrö, Saksvik and Torvatn, 1998; Parker, 2002, pp. ix–x, 57). Taking a broad conception of knowledge as all that a person knows or believes to be true about a particular subject, including his/her personal stock of information, skills, experiences and beliefs about the subject (Alexander 1991), the chapter examines how those involved in design and construction activities learned about safety matters, and the implications for firm performance for substantive safety outcomes and whether they comply with regulatory goals.

The chapter begins with an overview of a specialist body of knowledge, originating in the human factors and safety engineering disciplines, which is especially relevant to the safe design and construction of machinery. It finds little evidence that individuals involved in machinery design and construction drew upon this body of knowledge, unless their firms employed or engaged professionals who accessed and sustained the application of specialist information and methods. Instead, and consistent with a social constructivist perspective of learning (Billett, 1996; Palincsar, 1998; Vygotsky, 1978), individuals learned about safety matters through participation in design and construction activities, in the operations of their firms and interactions with external actors. In this respect, learning was situated as individuals constructed knowledge out of a wide range of materials, circumstances, and the histories and experiences of the people involved (on situated learning see Brown and Duguid, 1991; Brown, Collins and Duguid, 1989; Gherardi, 2008; Lave and Wenger, 1990). In addition, their personal histories and capacities, including the domains of knowledge and practices of their professions and vocations, provided different foundations for interpreting their experiences and constructing knowledge (Billett, 2001; 2003; 2006; Eraut, 2010).

The analysis presented here reveals that three practices were central to how those involved in machinery design and construction learned about safety matters – drawing upon their own and others' experience, interacting with their customers and referring to technical standards. However, actual practice, and safety knowledge constructed through practice, differed due to situational manifestations of practice in the operations of particular firms, and differences in the professional and vocational backgrounds of the principal decision makers in machinery design and construction (the key individuals), and the personnel engaged in designing machinery (the designers). Some of these practices and individual factors sustained better performance by firms for substantive safety outcomes, provided that individuals had access to learning rich experiences, while others were linked with poorer performance. Differences in firms' practices and individual factors also underlay the better performance of larger firms compared with smaller ones.

In bringing the separate scholarly fields of human factors and safety engineering, and professional and vocational learning, in conjunction with empirical findings about learning through participation in work activities, the chapter enriches understanding, across these fields, of the complex contextual factors that shape business capacity to manage risks. Also brought into focus is the influence of non-state institutions and actors, as firms' customers, suppliers and other industry contacts, as well as technical standards bodies. These insights into learning about safety in the work context are of interest to regulators, policy makers, specialists and practitioners seeking to build business capacity to manage risks and, in particular, to develop knowledge and skills to support safe design. In elaborating learning through practice as a central theme in explaining business performance, the chapter will also have wider resonance for regulatory, socio-legal and sociology of work scholarship. The findings about experiential learning challenge us to think beyond specialist and regulatory sources, to the everyday activities and interactions that shape business conduct and, as canvassed in the chapter conclusion, the implications for building capacity to achieve regulatory goals.

## Specialist Body of Knowledge for Safe Design and Construction

There is a considerable body of knowledge, originating in the disciplines of human factors and safety engineering, which addresses the integration of safety in the design of machinery and equipment, production systems, work environments and tasks. This body of knowledge is published in specialist texts and journals, as well as more user-friendly sources intended to facilitate access to this specialist knowledge. It comprises specialist information, methods and tools which, if they engaged with it, would set manufacturers up well for the structured analysis and resolution of safety problems in machinery design and construction.

The specialist body of knowledge provides fundamental information about the different ways that machinery may be hazardous, as well as different types of risk control measures and their effectiveness for eliminating and minimizing risks (Brauer, 1994; 2006; Corlett and Clark, 1995; Haddon, 1973; 1974; 1980;

Karwowski and Marras, 1999; Manuele, 1999b; Sanders and McCormick, 1993). It emphasizes and provides methods for designers to address safety matters from the earliest stages of design, in order to develop inherently safe designs and ensure that risk control measures are compatible with the functionality of designed items (Kletz, 1998a; Polet, Vanderhaegen and Amalberti, 2003; Reunanen, 1993, p. 108; Sagot, Gouin and Gomes, 2003; Seim and Broberg, 2010; Swuste, van Drimmelen and Burdorf, 1997). If such action is not taken, risk control measures may be removed or disarmed because they are a hindrance. The specialist body of knowledge also emphasizes the need for safety information to be well written, presented and structured, taking account of the target audience's knowledge and literacy, in order to enable end users to easily locate, read and understand such information (Corlett and Clark, 1995, pp. 87–88, 113; Reunanen, 1993, app 1).

This body of knowledge recommends that designers collect detailed information early in the design process about a wide range of factors, and conduct a holistic and multidimensional evaluation (Hasan, et al., 2003; Mills, 2000; Nachreiner, Nickel and Meyer, 2006; Paquet and Lin, 2003; Ringelberg and Voskamp, 1996, pp. 16–19). These factors include work tasks and processes across the life cycle of a designed item; the materials and products to be manufactured or handled using the item; the system in which the item is to operate; and dangerous events and hazards that could arise in these different interactions with the item. Designers are also advised to use a combination of practices to consider safety matters since different approaches contribute fresh perspectives and facilitate recognition of a wider range of problems and solutions (Busby, 2003, pp. vii, 2; Green and Jordan, 1999; Stanton and Young, 1999; Stanton, et al., 2005). There are techniques for data collection, representing tasks or processes, analysing human–machine interactions, assessing mental workload, predicting human error, assessing operator knowledge and understanding, analysing and representing communication, decision making and other aspects of performance, and conducting user trials to assess usability and human error (Brewer and Hsiang, 2002; Green, Kanis and Vermeern, 1997; Green and Jordan, 1999; Karwowski and Marras, 1999; Stanton and Young, 1999; Stanton, et al., 2005, pp. 6–12).

The specialist body of knowledge emphasizes the importance of designers involving end users and others with operational knowledge, in order to obtain information about user characteristics, and experience of the real nature of work activities, problems that may arise, and the potential for unintended use (Broberg, 1997; Garrigou, et al., 1995; Green, Kanis and Vermeeren, 1997; Morris, Wilson and Koukoulaki, 2004, p. 28; Raafat and Simpson, 2001; Reason, 1990, p. 9; 1997, 171–6). End users may interact with items in unintended ways for a wide variety of reasons including fatigue, work pressures, machinery faults, or to reduce strain and effort, save time or increase productivity. The objective therefore is to design so as to reduce the potential for unintended use, and to minimize the risk if unintended use does occur (Busby, 2003, pp. 12–14, 26; Fadier, de la Garza and Didelot, 2003; Fadier and de la Garza, 2006; Garrigou, et al., 1995; Kanis, 1998; Neboit, 2003; Polet, Vanderhaegen and Amalberti, 2003; Sagot, Gouin and Gomes, 2003; Weegels and Kanis, 2000). Available techniques to anticipate end user interactions with machinery and to stimulate discussion with users about problems and solutions include conducting user trials with prototypes, observing or video recording workers performing tasks with actual machinery or models, conducting scenario-driven discussions, and using virtual reality computer graphics (Brewer and Hsiang, 2002; Bruseberg and McDonagh–Philp, 2002; Butters and Dixon, 1998; Green, Kanis and Vermeeren, 1997; Haines and Wilson, 1998, pp. 17–26; Kuorinka, 1997; Paquet and Lin, 2003; Suri and Marsh, 2000).

Particular strategies may be needed to facilitate end user participation and communication with designers since end users may have difficulty identifying and articulating issues, especially if the new design is very different from their past experience (Bruseberg and McDonagh–Philp, 2002; Fadier and de la Garza, 2006; Garrigou, et al., 1995; Morris, Wilson and Koukoulaki, 2004, p. 29). There are different traditions for integrating human factors in design processes and they make different assumptions about the role of management, workers and other actors, the relations between these actors, and the methods used for participation (Jensen 2002a; Haines and Wilson, 1998, pp. 3–5, 47–56; Morris, Wilson and Koukoulaki, 2004, pp. 27–34, 76–87; Kuorinka, 1997). The various traditions imply that what constitutes effective participation by workers is contested and may require the support of management for worker participation, the development of workers' capacity to participate and good communication channels.

If those involved in machinery design and construction engaged with this specialist body of knowledge it could help them to appreciate the diversity of machinery hazards and the effectiveness of different types of risk control measures. It could increase their awareness that risk control measures are less likely to hinder the functionality of machinery if they are incorporated early in the design and are integral to the design of items. They might better appreciate that to be read and understood by end users, machinery safety information must be well written, presented and structured. They might understand that end users interact and struggle with machinery in unintended ways, for a wide range of reasons, and that it is crucial to anticipate unintended use and design to minimize this potential. They might also recognize that producing inherently safe machinery requires effective participation by end users, and that this may require the use of particular techniques, as well as the support of end use organizations, to facilitate user participation.

In principle, those involved in machinery design and construction could have accessed this specialist body of knowledge in safety engineering and human factors texts and journals, or through professionals working in the field of human factors and ergonomics, safety engineering or providing safety advice more generally. They could also have made use of software, tools and other products incorporating human factors and safety engineering techniques developed to make it easier for designers to apply specialist methods and reduce the perceived cost of doing so (Benedyk and Minister, 1998; Broberg, 1997; Feyen, et al., 2000; Gilad and Reuven, 1997; Jensen, 2002a; Schupp, et al., 2006).

There was, however, little evidence that study firms engaged with this specialist body of knowledge. Collectively, there were 24 firms in which those involved in design and construction activities had access to specialist information sources, advice from safety advisers or consultants, or both (36 per cent (24/66)). Six of these firms provided their staff with access to specialist texts, databases or other information resources (9 per cent (6/66)). For example, one firm had a handbook on machinery guarding, another had some books on ergonomics, and another had access to a database providing anthropometric criteria for designing machinery to be compatible with human body dimensions. As well as there being only a small number of firms using some form of specialist information, the sources these firms used were limited in scope when compared with the array of information sources, tools and methods offered in the specialist body of knowledge.

With regard to safety advisers, nine firms employed an in-house adviser who assisted with identifying potential sources of harm or determining control measures for the machinery produced by the firm (14 per cent (9/66)). These advisers dealt with safety matters generally in their firms, and their role included the safety aspects of the machinery designed and constructed. For example, the manager of a firm that manufactured materials handling and processing machinery explained the role played by the firm's safety trained tradesman:

> With the fellow that we do have here, his knowledge is really essential for us and his training, he chases up all the information that we need at this stage but if we didn't have somebody like that ... you need more help with it ... the information's out there if you've got the time to find where to get it all from. (Manager, Manufacturer 23)

Among these nine manufacturers with an in-house safety adviser, four had advisers with professional qualifications in safety who were responsible for managing safety in their operations. The other five firms had advisers who were managers or tradesmen with basic safety training. In these five firms, the role of the in-house safety advisers and the information they accessed were limited in scope, although in three of the five firms the safety advisers were at least the reason their respective firms engaged with the provisions of Australian OHS law applicable to machinery design and construction.

Nine firms engaged consultants to advise them on particular aspects of machinery safety such as mechanical hazards or guarding, ergonomics or vibration (14 per cent (9/66)). For example, a manufacturer of machinery used in the production of vehicle parts contracted consultant physiotherapists to provide advice on the ergonomic aspects of machinery design. An engineer with the firm explained:

> Usually if we have something that is really awkward we have access to the physio people and we say look this is what we want to do and they're quite happy to sit down with us and say "well, yeah that's acceptable" or "no, it's not" ... [They] give us a tick or tell us to change it ... They're not in-house ... we can call on them through our occ health and safety nurse as needed. (Engineer, Manufacturer 11)

This manufacturer and the other firms that used consultants engaged them when they identified the need for specific assistance. Other than for advice about machinery hazards or risk control measures, eight firms engaged consultants to conduct a risk assessment for their machinery (12 per cent (8/66)). Through their involvement in risk assessment, these consultants also contributed to decisions about hazard recognition, risk control and provision of safety information. (Risk assessment is examined further in Chapter 7).

Overall, the use of specialist information and general safety advice did not feature strongly in the activities of study firms. Among the 24 firms that accessed specialist information sources, had an internal safety adviser and/or engaged external consultants, only three firms accessed specialist information or advice in a more organized and systematic way (4 per cent (3/66)). The other 21 firms accessed such information or advice spasmodically, on an ad hoc or piecemeal basis (32 per cent (21/66)). For all 24 firms, the information accessed, and the role of in-house advisers and external consultants were quite limited when compared with the sources, tools and methods provided and recommended in the specialist body of knowledge. The remaining 42 firms did not access published information other than technical standards or supplier information, as discussed below, did not have an in-house safety adviser, and did not engage external consultants for advice about machinery safety or to conduct a risk assessment (64 per cent (42/66)).

Although a wide range of specialist information, methods and tools were available to support the integration of safety in design, study firms made little use of these specialist resources. Was this simply a failure by firms to seek out information and engage with available resources? The explanation is not so one sided. A key reason for the lack of uptake and application of specialist sources, methods and tools was the nature of the specialist body of knowledge. Particular approaches have typically been developed and applied by human factors or other safety professionals, who also support and facilitate their application to specific designed items or settings (Badham and Ehn, 2000; Hale, Kirwan and Kjellen, 2007; Jensen, 2002a; Haines and Wilson, 1998, pp. 52–3; Sagot, Gouin and Gomes, 2003; Seim and Broberg, 2010). In other words, professionals act as a conduit to specialist information, methods and tools, and sustain the application of this body of knowledge.

Only one manufacturer in the present research employed an ergonomist and another engaged consultant physiotherapists to provide ergonomic advice. These two firms were the only ones that also had some ergonomics references available to their designers. There was no evidence that the consultants engaged by other firms, to provide ad hoc advice on aspects of machinery safety or to conduct risk assessments, contributed insights from the specialist body of knowledge or made this knowledge more accessible to those involved in machinery design and construction.

While the specialist body of knowledge offers valuable insights about integrating safety in the design and construction of machinery, in many firms those involved in these activities did not have access to or have past experience of this specialist body of knowledge. In firms that did not employ or engage human factors or other safety professionals, it was not part of their practice to make use of specialist information, methods or tools, even though there were resources available to make it easier for designers to access and apply these. The practice of machinery design and construction was typically disconnected from the specialist body of knowledge.

## The Wider Practice of Machinery Design and Construction

*Overview of Wider Practice*

Rather than drawing upon specialist sources, those involved in the activities of machinery design and construction learned about safety matters through participation in these activities. That is, they learned about safety in the everyday operations of firms and interactions with external actors. Based on the accounts of interviewees in 66 firms, typical activities in design and construction were gathering and analysing information, identifying design requirements, generating plans, manufacturing or purchasing and assembling components, and trialling or testing machinery, or components of it. In the course of these activities, those involved considered machinery safety problems and determined what action to take, along with a series of matters such as machinery operation and functionality, power consumption, quality, cost, and customer acceptance of machinery. They identified problems or areas for improvement for all of these matters, and progressively refined or modified the design of the machinery and its construction.

Three aspects of machinery design and construction reflected wider practice and, again based on interviewees' accounts, these three aspects were the principal ways of learning about safety matters. The three central elements were: (1) drawing upon experience producing machinery (within and outside the firm); (2) interacting with customers; and (3) referring to technical standards.

*Drawing upon Experience*

Much of what those involved in design and construction knew about machinery safety was derived from experience. This was ontogenetic knowledge (Billett 2001, 2003); that is, knowledge derived through individuals participating in design and construction activities in their current firm and other firms, and through contact with other producers of machinery and industry sources more widely, across their working lives. It was experience as "the amalgam of fact and fiction which make up the normal human repertoire" (Green, Kanis and Vermeeren, 1997).

The key individuals in most study firms had worked for their current firm for ten or more years (62 per cent (41/66)), or had long term experience with the same type of machinery and the work processes in which it was used (17 per cent (11 of 66)). Their knowledge of safety problems and solutions was derived from experience with the same or similar machinery, and the way that machinery was constructed to deal with particular problems. As explained by two key individuals who had respectively worked with their current firms for 28 and 33 years:

> We tend to retain people for an enormous amount of time ... and we've often got people in the 20s to 30s [years of service], which has been almost the history of the company. So a lot of the knowledge is passed on through the different groups or drawings that we have, technical drawings nominate the standards that we have to build every machine ... if four or five people left, the machines wouldn't change but I would say most of the knowledge we've learnt it's become common sense and it's part of our culture rather than having a magical book here to say how a machine should be built. (Managing director, Manufacturer 5)

> ... because we've been making the same types of products for a good many years and within the company probably for 50 years, then you have a certain amount of knowledge and experience. ... I grew up through that field as a technical manager in installations, I know how the equipment is used and I can service it and I can install it, fine tune it, run it, commission it and you get to know more about the people that use it. (Manager, Manufacturer 26)

These two examples are typical of the long term experience of those involved in machinery design and construction. They borrowed and patched together various pieces of their past experiences (Gergen 1994, pp. 269–70). They also learned about machinery safety matters through contact with other firms. As the managing director of a firm that manufactured timber handling machinery stated:

> ... because we're agents for the US companies, the Swedish companies, the German companies, the French companies, we get a lot of experience from that; hundreds of years of experience. With that sort of base, we have a lot of experts and if something is occurring we can actually ring them and say we've got this occurring with a machine, what do you suggest we do? ... So it's not limited to our staff ... We've got a very, very big base to draw from and we draw from it ... we can just fine tune the machine and alter timing and things like that, and it can make a big difference to the performance of that machine, and the safety of that machine. So we look at ourselves as world wide and not just local here, Australia. (Managing director, Manufacturer 21)

Similarly, the owner of a firm that manufactured agricultural machinery stated:

> I suppose I pick up things from experience. I exhibit at the world's largest field day Tulare in California. I travel ... extensively through places like Stuttgart, Bordeaux, Montpelier ... go to the Royal Show at Kenilworth in England. I see a lot of stuff and I think "mm, that's a good idea", or whatever. And I take a lot of photos, I collect a lot of brochures ... They're the things that I look at. (Owner, Manufacturer 19)

These examples illustrate how those involved in machinery design and construction drew information and ideas from other companies in their networks, as well as trade shows and field days displaying their type of machinery. This finding resonates with European research which suggests that designers, in particular engineers, are intuitive, work from experience, find analogies with other known examples, and prefer dialogue with others in their own company or other designers (Broberg, 1997, 2007; Hale and Swuste, 1997; Swuste, Hale and Zimmerman, 1997). They favour solutions previously proven to be effective, and consolidate their knowledge and resolve safety problems through experience and interactions with others. For those involved in machinery design and construction, learning through experience in their current firm and in other firms, and from their industry contacts more widely, was a way to become familiar with state-of-the-art safety features, or at least industry standard measures, for their type of machinery (see also Main, 1999; Standards Australia, 1996, p. 38). It was also a way for them to learn methods for machinery risk assessment, as examined in Chapter 7.

*Interacting with Customers*

Also central to everyday practice in machinery design and construction were interactions between manufacturers and their customers, as procurers, distributors and sometimes end users. Most study firms sought or received some form of input from their customers (89 per cent (59/66)). For example, the owner/engineering manager of a firm that manufactured food processing machinery explained:

> I think listening to the customers is probably the best because they've often got very big factories, they've got hundreds of plants operating ... And it'll be their requirements that they ask you to incorporate that will make you think. ... they've got that experience and they're looking for ways in which problems can be eliminated. ... they don't tend to supply that to you in a documented form, but they will pass on perhaps how they've done it in the past and how they see it should be done ... . (Owner/engineering manager, Manufacturer 49)

Customers were a source of advice about machinery safety problems, and alerted manufacturers to issues which they might otherwise overlook. Customers also provided solutions to problems and influenced firms' choice of action, including methods for risk assessment (see Chapter 7).

There were, however, wider economic motivations for manufacturers and their customers to interact, and both parties sought and took opportunities for this to occur. For manufacturers, such interaction was a way to share responsibility for the end product, to build relationships that were important for securing repeat business, to be made aware of problems that could damage the firm's reputation, and to ensure that machinery was acceptable and competitive in the market place with regard to quality, functionality or cost. From the customer's perspective, interaction with the manufacturer was a way to ensure machinery was acceptable before it was delivered, and avoid difficult or costly negotiations or legal action, which could ensue if problems were not resolved before the customer took possession of the machinery. On the other hand if problems arose after the machinery was supplied, there was an incentive for the customer to report them to the manufacturer and try to ensure the latter took responsibility for resolving them.

There were therefore a variety of reasons for interaction between manufacturers and their customers, and such interaction was common practice. The safety of machinery was only one of a series of concerns addressed through manufacturers' practice of seeking and receiving input from customers but it was also integral to these concerns in the sense that, for example, a firm's reputation might be damaged or costs incurred by unresolved safety problems. Manufacturers' interactions with customers were manifest in many different ways, as demonstrated later in this chapter.

*Referring to Technical Standards*

The only published information sources commonly referred to in study firms were technical standards, which provided more detailed specifications and procedures. Most firms referred to technical standards in the course of machinery design and construction (76 per cent (50/66)). To the extent that they applied one of the standards that was mandatory or evidentiary under Australian OHS law they were taking steps to comply with the law, but the key individuals interviewed were generally not aware of the legal status of particular technical standards (see also Chapter 4). Technical standards and the non-state institutions promulgating them were regulatory influences, whether or not particular standards constituted legal rules.

Manufacturers used technical standards for several key reasons. They valued these standards as a source of specific information as, for example, observed by the owner and managing director of a firm that manufactured agricultural and other machinery:

> ... there's things like welding standards and construction things ... when you're doing a weld it should be this long or you know have this much penetration and all those sorts of things ... and there are things that relate to sub-componentry of a lot of what we do, wiring for instance and things like that. (Owner/managing director, Manufacturer 68)

Technical standards not only provided information about safety matters, they also provided specifications for fitness for purpose, quality, and public health and environmental matters (Productivity Commission, 2006, pp. 6, 14). Such standards were therefore relevant to manufacturers' wider economic motivations. Technical standards also provided benchmarks of accepted industry practice which applied Australia-wide (Australian Standards), or globally (international and European harmonized standards). Manufacturers believed that they helped to establish a level playing field in the markets for their machinery, which was an important concern for some firms. (For further discussion of competitive pressure and motivations see Chapter 8).

Manufacturers used technical standards specific to their machinery such as standards for conveyors, robots, cranes and hoists, or boilers and pressure vessels. They also drew upon standards with more general application to machinery and structures, for example those for safeguarding of machinery, electrical safety, platforms and walkways, and structural steelwork. For instance, a manufacturer of cranes and hoists applied a series of Australian Standards specific to the firm's machinery, and some with more general application. The production manager explained:

> Well we're governed by a couple of things. I can't remember the crane code, it's 1418 from memory. There's another one AS 1554, which is relating to structural steel, so we have to abide by that. What else do we use? SAA wiring rules, which is another one which governs how it's wired. (Production manager, Manufacturer 47)

As well as information relevant to the design and construction of machinery, some technical standards provided methods for assessing and managing risks, and specialist methods for risk analysis (Standards Australia 1996; 1998; 1999; subsequent editions are ISO 2009; Standards Australia 2006b). Technical standards were therefore constituents of procedural knowledge, as well as content knowledge about types of hazards, risk control measures and other matters (for types of knowledge see Alexander, 1991).

While some interviewees considered that technical standards were costly, difficult to interpret at times, or that particular provisions of standards could not be applied for technical reasons, most manufacturers used technical standards as an information source. They retained print or electronic copies, or accessed them through a subscription service. They drew upon specific standards to the extent that they provided useful guidance, and departed from them if they perceived that certain provisions were not applicable or that there was a better alternative. As one engineering manager stated, "there are aspects of them we can use directly and others we have to interpret because nobody writes the standards specifically for our equipment" (Engineering manager, Manufacturer 55). Several firms had key individuals who were members of standards development committees for the type of machinery they produced (see also Productivity Commission, 2006, p. 268). These firms not only applied particular standards in designing and constructing their machinery, they also influenced the content of those standards.

The prominence of technical standards as an information source for machinery design and construction has also been recognized in European research. A study of agricultural machinery design in the UK found that design engineers used the European harmonized standards or British (BSI) standards more than any other readily available information source (Crabb, 2000, pp. 26–7). A second study, focusing on the design process for production plant in a large Danish firm, revealed that technical standards were ranked highly by design engineers as an information source and for problem solving (Broberg, 1997). Like the practice of learning from the experience of other machinery producers, referring to technical standards helped designers to avoid re-inventing solutions and provided an indication of accepted industry practice (see also Hunter, 1999; Kerwer, 2005; Main, 1999).

*Summary of Wider Practice*

Three practices were central to how those involved in design and construction activities learned about machinery safety matters. They drew upon their own and others' experience as producers of machinery, interacted with their customers (as procurers, distributors or end users), and referred to technical standards.

Other safety and socio-legal researchers have recognized that organizations may learn about health and safety matters, and how to comply with social and business regulation more generally, through their customers or other organizations with which they have business relationships (Gunningham and Sinclair, 2002, pp. 17–18; Hopkins and Hogan, 1998; Hutter, 2011, pp. 90–4, 96–8; Lamm and Walters, 2004; Walters 2001, pp. 52, 375; 2002, pp. 45–6). It is also well established that designers learn through experience and their interactions with others, and have a preference for technical standards (Broberg, 1997; 2007; Hale and Swuste, 1997; Swuste, Hale and Zimmerman, 1997). The present research has established that experience, customers and technical standards were not only favoured but were the principal constituents of knowledge about machinery safety for those involved in its design and construction. It highlights the significance of non-state actors and institutions, as parties in supply chains and networks, and technical standards bodies, as sources of information contributing to manufacturers' knowledge base.

Whether safety knowledge constructed through these central practices or through other sources was sound, was contingent upon situational manifestations of practice in the operations of particular firms. It was also influenced by the personal histories and capacities of the individuals involved. In other words, knowledge construction had different starting points and constituents. These are the issues explored in the next section.

## Individual and Situational Differences

*Overview of Individual and Situational Differences*

While it was common practice for those involved in machinery design and construction to draw upon their own and others' experience, to interact with customers and to refer to technical standards, there were multiple bases from which individuals constructed knowledge about safety matters. Within and additional to the three central practices there were situational differences in the actual practice of design and construction in the context of each firm's operations. Actual practice, and safety knowledge constructed through practice, varied as operational activities, technologies and the specific nature of interactions with external actors differed between firms. Individuals involved in design and manufacture activities also constructed safety knowledge from diverse bases due to their different personal histories and capacities. This section elaborates the individual and situational differences that shaped safety knowledge, once again drawing on the accounts of interviewees in the 66 study firms.

*Individual Differences in Personal Histories and Capacities*

One area of difference was the diverse professional and vocational backgrounds of the key individuals and others involved in the design of machinery in study firms. These differences meant that key individuals and designers drew upon various sociocultural traditions, and the domains of knowledge, practices, values, technologies and norms of those professions or vocations (Alexander, 1991; Billett, 2001; 2003; 2008a). Differences in occupational background also meant that key individuals and designers had different personal histories and capacities from which to interpret their experiences in machinery design and construction. They had different foundations for their learning.

The key individuals in some firms were engineers (33 per cent (22/66)). Most commonly they were mechanical engineers but there were also electrical, manufacturing, electronic, structural and chemical engineers. In other firms the key individuals had a degree or diploma in a different field (15 per cent (10/66)). Their qualifications were wide ranging and included business administration, economics, computing, science, agricultural science, farm management, marketing, arts and education. Also, some key individuals were tradespeople (35 per cent (23/66)). There were technical draftsmen, boilermakers, electricians, fitters and turners, mechanics and welders. In the remaining firms the key individuals had no professional qualification or trade background (17 per cent (11/66)). In addition, whether or not a firm's key individuals were engineers, many firms employed or engaged engineers as designers (70 per cent (46/66)). In the other firms, designers either had a trade background, or they had experience constructing machinery but no qualification or trade (30 per cent (20/66)).

A further way in which the personal histories of key individuals and designers differed was for operational experience. In two firms the key individuals and designers had previously worked in industry using the type of machinery they now produced (3 per cent (2/66)). They contributed an end user perspective to the design and construction of their machinery. For example, the director of one of these firms which made food processing machinery, explained:

> When we design the equipment, we design it from the point of view of the user
> and all the people in the executive of the company have many years of experience
> in the using side of it, and we came out of industry and now we work in making
> equipment for industry ... and we've all been there and used the equipment, so

we're very conscious about the fact of safety and we don't consciously put out anything that we don't believe is safe, so we include those sort of perspectives when we're designing it deliberately. (Director, Manufacturer 42)

The fact that key individuals or designers had operational experience, or a particular professional or vocational background, does not imply that they had knowledge relevant to machinery safety matters. To the extent that safety has been addressed in professional and vocational education programmes in Australia, the focus has mainly been on the roles of employers and workers, rather than designers and manufacturers, especially at the time of data collection for the research (Caple, 2000; Toft, Howard and Jorgensen, 2003; and see ASCC, 2008; NOHSC, 1998). The different backgrounds simply meant that key individuals and designers constructed knowledge from different bases. For example, engineers might have knowledge of structural loads, hazardous energies or machine dynamics (see for example SUT, 2009). On the other hand, key individuals and designers with years of experience operating or maintaining a particular type of machinery in industry would be better placed to understand the real nature of work with that machinery and the safety problems arising in end use.

In essence the constituents of safety knowledge, and how they were interpreted and laid down as knowledge, differed according to the personal histories and capacities of the individuals involved in machinery design and construction. Their professional or vocational backgrounds, and past operational experience, constituted alternative bases and pathways for learning about machinery safety matters in the course of their current work roles.

*Situational Differences in Standards, and Design and Construction Practice*

Over and above diverse individual foundations for constructing knowledge about machinery safety, situational differences in firms' practices gave rise to multiple constituents of learning. These situational differences related to the markets for particular machinery, the type of technology used in design and the type of machinery produced. There were also differences in the methods used to identify safety and other problems arising from machinery (functionality, quality and so on), ways of determining how to resolve problems and approaches to trialling or testing machinery. This diversity engendered multiple sources of information, perspectives and understandings of safety matters.

By virtue of different markets for their machinery, some manufacturers sourced additional information in the form of technical standards relevant to particular markets. Firms that exported, or intended to export, their machinery overseas used international (ISO) or European (EU) harmonized standards (14 per cent (9/66)), while still also making use of Australian Standards. These exporting firms perceived that the ISO or EU standards provided an indication of acceptable standards for their machinery in European and other overseas markets. The EU harmonized standards were clearly relevant to firms supplying machinery into

Europe (European Commission, 1998a; 1998b; 1998c; see also Chapter 2), and some firms exporting machinery to China or other Asian countries believed that the European harmonized standards also had standing in that part of the world. As the engineering manager of a firm with markets in Europe and Asia stated:

> ... we call on European standards as well ... as kind of, I suppose, the parent standards internationally, and Australia seems to be evolving towards the European standards. So by using the European we tend to cover other countries. (Engineering manager, Manufacturer 55)

A different example of a manufacturer using standards for particular markets was a firm that supplied machinery into the international maritime industry. This firm applied technical standards developed by classification societies, the audit bodies operating in the maritime industry, because these standards specified acceptable criteria for the design and construction of machinery supplied into this industry (International Association of Classification Societies, 2006).

Apart from different technical standards, some manufacturers used design technologies which afforded alternative perspectives of their machinery, its components, and potential problems and solutions. For instance, it was common practice for firms to use computer aided design (CAD) software to produce two or three dimensional models of their machinery and, in about one quarter of firms, designers took steps to eliminate hazards or otherwise resolve safety problems during CAD (23 per cent (15/66)). An example was a manufacturer of materials handling and special purpose machinery, which used CAD software incorporating a mannequin to assess the interaction of the machinery with operators, vehicles and other aspects of the workplace. The managing director demonstrated this on his computer for a packing line he was designing, explaining:

> Typically in any instance we will try and look at how the operator is working. He's way too close to that motor ... we need to provide access ways so therefore that access way must be gated ... that access way is perfectly safe, it's back to return [closed], but these are things that we bring up at this stage ... we had the layout previously, presented the operator's risks, the forklifts, we still had an issue here with the operator with the forklift so we bollarded it. (Managing director, Manufacturer 1)

By using CAD technology, designers could bring together the structure and component parts of machinery, and locate the position of operators. Some programmes allowed designers to produce a working model, make parts move, view them from different angles and see what different parts might hit or obstruct. With three dimensional versions it was easier for designers to visualize heights, distances, work postures, functions of the machinery and the work environment (Sundin and Medbo, 2003; and see Pappas, et al., 2007). These perspectives were only available to firms with the resources to purchase the technology and time to train staff in its use, or the resources to purchase this capability as an outsourced service. In particular, more complex three dimensional forms of CAD software were more costly and required intensive training and experienced personnel who worked with the software routinely (Ottosson, 2002; Sundin and Medbo, 2003).

Many manufacturers used simpler approaches to identify problems and determine solutions as they designed and constructed their machinery. The most basic approach was visual inspection of machinery, which was used by half of the study firms (50 per cent (33/66)). Typically, those involved in design and construction activities simply looked at new machinery as it was built, or inspected existing machinery of the same or a similar type, walked around it and viewed it from different angles. As interviewees variously explained:

> ... basically looking at it and literally not being able to get your foot under there, me put my foot under there and spin the blade, no I can't get to it that's fine. (Managing Director, Manufacturer 3)

> ... what we generally do is just get a couple of us engineers together and walk around a machine. (Engineer, Manufacturer 13)

> ... we just keep looking, looking, looking and whenever you see a deficiency fix it and that's from the start of the process to the finish of the process. (Safety manager, Manufacturer 18).

Inspection was a way to look for aspects of the machinery that could hurt people such as whether someone could get their fingers or other body parts into a chain, auger or other danger zone. It was also a way to identify opportunities to improve machinery and determine solutions to problems, such as how to prevent or reduce the potential for access to dangerous parts.

Another type of inspection focused on the layout and conditions of the work environment into which the machinery would be installed or used. Comparatively few firms conducted such an inspection (17 per cent (11/66)). The manager of a firm that manufactured fruit processing machinery explained the approach:

> You actually travel to the site and see where are they going to operate their forklifts from, where are the storage areas, where are the cool rooms, switchboards, catwalks, all of those type of things ... in principle that's what you do, you look at where it's used, how it's used, how it's applied and the environment around it and you figure out as best you can what the hazards are. (Manager, Manufacturer 26)

Inspection of the end use work environment was an important method because hazards and hazardous situations arose not only from machinery itself but also from its interactions with people, and other machinery and features of the work situation in which it is used. The manufacturers that used this approach gained insights into how their machinery would interface with operators and others in the vicinity of the machinery, as well as other machinery in the workplace, including moving vehicles. They also considered the location of their machinery in relation to fixtures and fittings such as fire escapes or fire fighting equipment, and physical conditions such as stability or potentially corrosive substances on the ground or surface on which their machinery would be installed on.

Moving beyond technical standards, technology and inspections, some manufacturers became aware of safety problems with their machinery through information received or obtained from customers or other sources about injuries, incidents or hazardous situations involving their machinery, or similar machinery produced by another firm. Just less than half of the study firms were alerted to particular safety problems in this way (45 per cent (30/66)). For example, a manufacturer of special purpose machinery obtained such information from several sources. The business development manager explained:

> Sometimes if we're redesigning a machine that they've [the customer] had in existence we'll ask for their safety data on that in terms of any incidents, near misses or whatever ... apart from some of the blue chip companies, a lot of companies just don't keep the near miss information and we're trying to encourage that, so we get that type of data. We do go to AIG (an industry association] and to WorkSafe [the OHS regulator] and say to them, you know, "this particular type of bandsaw is there any data there that we can get access to?" And AIG has a health and safety section that can give general analysis. (Business development manager, Manufacturer 56)

This manufacturer actively sought information about machinery injuries and incidents from customers and other sources. Some other firms only received this type of information reactively, if their customers reported injuries, incidents or hazardous situations with the firm's machinery.

Another information source was suppliers of general or safety-related components. About one quarter of study firms received assistance from suppliers in resolving safety problems (27 per cent (18/66)). This might simply involve the manufacturer receiving product information from sales representatives. For example, the managing director of a firm that manufactured metal processing plant stated:

> There's a certain amount of pressure which comes from our supply chain. "We've got these new safety devices". And it's not so much pressure as opportunity. And we say, "that's a good idea, we'll do that". (Managing director, Manufacturer 40)

Other firms actively sought assistance from suppliers with the resolution of particular safety problems. This practice was illustrated by a manufacturer of agricultural machinery, which asked a supplier of hydraulic components to help solve a manual handling problem. The issue impacted upon the firm's workforce in production and, if unresolved, would have affected workers maintaining the machinery in end use. The production manager explained:

> We buy in things called hydraulic hose kits so the machine has a kit and at the moment we're building big machines so the hydraulic hose kit comes in with four packs of hoses in one plastic sheet, which a person cannot lift ... so we've just had [the supplier] up to have a look at them ... These would have to be 60 kg ... so now we'll work hard with [the supplier] on working out a solution around this and we have to go back to [the designer] to say well the kits are fantastic but there's a problem because I can't lift them out of the boxes that they come in ... the guys on the line just go and pull a cart out and take the hoses out of the cart. I can't lift the stuff into the carts so that's an issue [the designer] is going to have to look at and along with [the supplier]. (Production manager, Manufacturer 29)

Manufacturers sought assistance from suppliers either because, as suppliers of safety products, they offered practical assistance and solutions, or because they could produce component parts to meet particular safety needs. In some cases suppliers had specialized, technical knowledge about safeguards such as interlocking guarding systems or electrical safety devices. Other research on how designers work has similarly shown that they commonly discuss problems with suppliers or obtain literature from them, and safety research more generally has observed that suppliers are often part of firms' trusted networks and valued as sources of know-how and capability on safety matters (Broberg, 1997; Boston, Culley and McMahon, 1999; Walters, 2001, p. 52).

Further situational differences in design and construction practice related to the type of machinery. For manufacturers producing machinery supplied in larger numbers, trialling a prototype or model was an essential process when developing new designs or making changes to existing ones, before going into full scale production. Half of the study firms identified potential sources of harm or determined risk control measures in the course of trials using prototypes or models (50 per cent (33/66)). Typically, firms built an item of machinery as a prototype, but some prototyped particular parts of the machinery and some used a scale model made from cardboard or other common materials. In a cycle of trialling and rebuilding prototypes or models over a period of months or, in some cases years, manufacturers progressively refined the design before commencing larger scale production and supplying the machinery for wider use. They used prototypes or models in their own in-house trials of machinery, and some also involved customers to obtain their input. (User trials are discussed further in the section on customer interactions below).

The owner of an engineering workshop demonstrated his approach to examining machinery facilitated by a prototype model. He talked through the sources of harm he had identified with an earlier prototype for a machine he was developing. He drew upon his knowledge of how the machinery would work in order to identify how an operator could become trapped in it during operation. He explained how he had changed the design to remove a downward, crushing action by a piston because there was a risk of an operator being injured. He had reduced the crush risk by reversing the direction and speed of the piston so that it rose upwards slowly, and incorporated overhanging flanges and a sliding lid on the top of the machine to reduce the risk of the lid blowing off with the upward pressure. He had also fitted a limit switch to ensure that the machine could not be operated with the lid open, and used the prototype to test the structural integrity of the prototype by applying a pressure of several hundred kilograms.

In general, those involved in machinery design and construction used a combination of sources and methods to consider safety issues, but the approaches they used did not reflect the kinds of sources, methods and tools advocated in the specialist body of knowledge, as discussed above. Rather, their approaches were consistent with everyday practices for designing, constructing, trialling and testing machinery. It is also important to note that some manufacturers conducted a form of risk assessment (58 per cent (39/66)). As with other practices used by manufacturers, situational differences underlay whether and how firms conducted risk assessment (see Chapter 7).

Differences in markets, technology for design, the type of machinery, and the methods used by firms meant that those involved in design and construction activities drew upon different sources of information, and gained different perspectives of the safety problems arising from their machinery, and ways to resolve them. In one firm, those involved in design and construction only knew of safety problems through reports of injuries or incidents occurring with the firm's machinery in end use. In another firm, those involved in design and construction knew about safety problems and possible solutions by inspecting machinery as it was built, trialling prototypes, reviewing reports of injury or incidents with the machinery in end use, and referring to relevant Australian and international technical standards.

As a consequence of situational differences in design and construction practice, those involved in these activities had quite different bases for constructing knowledge about machinery safety matters. There is also a wider implication of these findings. The constituents of safety knowledge went beyond authoritative sources such as legal instruments, OHS regulators' guidance materials, or specialist sources. Indeed, use of these types of authoritative sources represented less common situational manifestations of practice in the operations of a small proportion of firms.

## Situational Differences in Interactions with Customers

Further situational differences arose from manufacturers' dealings with their customers. While most firms interacted with their customers on machinery safety matters (89 per cent (59/66)), these interactions were manifest in diverse ways as the parties initiating the interaction, the persons involved, the timing and the methods of interaction differed between firms.

With regard to the parties initiating interaction, almost two thirds of study firms actively sought input from their customers on machinery safety matters (62 per cent (41/66)). The remaining firms received input if their customers provided it (27 per cent (18/66)), or did not receive customer input at all (11 per cent (7/66)). In all of the 59 firms that sought or received some form of input, various parties contributed. This always included the customer's managers or engineers and, in addition, some exchanges involved end users with experience operating or maintaining the machinery, or working in processes in which the machinery would be used (38 per cent (25/66)), and/or personnel in a safety role as health and safety representatives, committee members or safety advisers (18 per cent (12/66)).

The potential contribution of each of the parties providing input differed. The manager/engineer group were more likely to be aware of the intended use for the machinery and the operations in which it would be used. End users could contribute practical insights about the way they interacted with the machinery and its impact upon them. Worker representatives, committee members and safety advisers were more likely to have some safety training, an appreciation of different types of hazards and control measures, and might also have skills in representation to assist in clarifying and raising issues identified by end users (see also Blewett, 2001; Vanderkruk, 1999; VTHC, 2004).

The stage at which firms sought or received input was another source of variation with implications for understanding safety issues. Some firms sought or received input at the design stage (59 per cent (39/66)). Others did not seek or receive input until the construction stage (6 per cent (4/66)); the time of installation or supply (5 per cent (3/66)); or when the machinery was in use in the workplace (18 per cent (13/66)). Input at the design stage could draw attention to ways to eliminate hazards or minimize risks that were integral to the design. On the other hand input about safety matters at a later stage could only highlight the need for safeguards that would be add-ons, or signal the need to redesign the machinery at some time in the future.

Manufacturers typically sought or received input from customers through a combination of methods, but there was considerable diversity in the methods used. Some firms reviewed design plans in consultation with customers (12 per cent (8/66)). These might be CAD or other drafting plans. For example, a manufacturer of materials handling machinery inspected the end use work environment, and used CAD plans or other visual methods in consultations with the customer's engineers and end users. The firm's business development manager explained:

> During the early design stages when I'm talking to the engineers I always like to get the operators involved to get their feedback ... If it doesn't suit them or it's cumbersome or something they will not use it ... So I see the operators as being one of the key cornerstones in those focus meetings to determine how we're going to go about it ... What we do is we go and look at an application ...

say for example, he's got a stillage full of engine components. He's got to lift them out, he's got to load them into a machining centre. Now because he's got to reach in and this thing weighs 20 kilos, he can't deal with it ... The whole idea is just to identify the movements of that operator and then we go away and we'll come back with some CAD drawings ... and then you talk to the operators and engineers and say well this is what we're proposing and explain to the operators that, using whether it be video footage, whether it be using photos, whether it be CAD drawings, that this is what it's going to look like and this is how you'll use it, do you feel comfortable with it? ... this is how we're going to do it and this is going to reduce the risk of you hurting your back, dropping a casting on your foot or getting tangled up in that ... . (Business development manager, Manufacturer 59)

For this manufacturer, observing and taking photographs or videoing work tasks was the basis for preparing CAD plans. The firm could then use all of these visual materials to facilitate consultation with the customer's engineers and end users.

Another way that some manufacturers obtained input was by conducting user trials of their machinery (26 per cent (17/66)). This involved giving a prototype, model or the finished product to a customer, distributor or end users to trial in the kind of operational setting(s) in which the machinery would ultimately be used. The item might be left with customers, distributors or users in workplaces for several months or longer, to enable them to thoroughly interact with it and then advise the manufacturer about any problems they had experienced. As the technical services manager of a firm that manufactured saws and other construction equipment explained:

We have a select group of consumers out there, contractors, and they do testing for us ... They're very critical. They're very honest and we take all of what they say on board because these guys make a living out of using this machinery ... and we rely on a lot of their experience and a lot of their use ... we can make a machine, that's not a problem, but can we make it run safe ... They look and think about a product from a totally different perspective because they're looking at it as a user. We're looking at it as a manufacturer, looking at it to sell. So we've had a few problems where we've thought, "yeah, this is going to be a goer" and it's just totally not in the ball park. (Technical services manager, Manufacturer 12)

As this example shows, conducting a user trial was a way for a manufacturer to access customers' experience. It also alerted the manufacturer to problems experienced by end users.

Rather than seeking input from their customers, some manufacturers responded to their customers' safety requirements. For example, some firms received detailed specifications for commissioned machinery and the customer checked that the manufacturer implemented these requirements (14 per cent (9/66)). Key actors in these activities were the customer's managers and engineers, and sometimes safety personnel, worker health and safety representatives or committee members. For example, a manufacturer of special purpose machinery received specifications and input from customers that were large businesses in the manufacturing industry. The owner of the firm explained:

> When it comes to special purpose built equipment, usually companies like [named three large customers], just to name a few, they've got their safety committees, and they've got their safety requirements and they are laid out in their specification. So when you're quoting, you're quoting actually the equipment applying to all those standards. For this equipment which is out there at the moment we had a member of the safety committee from [one large customer] here yesterday, going through the whole way of how the product is made, and then telling us ... which parts of the machine need to have things added or modified to comply to their standards. And when the equipment goes out to [that customer] then the safety committee will go over them with a fine comb. (Owner, Manufacturer 14)

This example illustrates how some customers took the lead in providing input and played an active role in ensuring their safety requirements were met. This might include the customer ensuring that the manufacturer understood safety requirements in specifications, explaining how the machinery would be used, reviewing the design before construction commenced, checking the machinery once it was produced, and/or advising the manufacturer of areas in need of improvement.

The most common method for manufacturers to receive input was for customers to provide ad hoc feedback about machinery, orally or in writing, after it was in use in workplaces (44 per cent (29/66)). Such feedback might be provided by phone, through warranty claims, by speaking to the manufacturer's representatives at field days or other casual communication. The response of one manufacturer of agricultural machinery was typical of this approach. The managing director of the firm stated:

> I have a one to one connection with my customers at field days which I'm doing all the time, so we would certainly find out those sorts of things if it was a problem. (Managing director, Manufacturer 20)

This statement exemplifies a typical assumption among the manufacturers that relied on their customers to provide feedback, that if there were any problems with the machinery, those experiencing problems would alert the manufacturer. Ad hoc feedback was the weakest form of input on machinery safety matters, if it was the only form of interaction between a manufacturer and their customers.

Other than discussion of design plans, user trials, customer specifications and ad hoc feedback, additional forms of input in study firms were dialogue-based consultation (35 per cent (23/66)); joint inspection of the finished or existing similar machinery, or inspection by the customer alone (11 per cent (7/66)); and reports of problems through a quality programme (11 per cent (7/66)). Methods used by five or fewer firms were organizing a special forum to obtain input from customers or distributors, viewing video or photographic recordings of the end use workplace or machinery, testing machinery jointly with customers or surveying the target audience to gather feedback. Finally, it should be noted that some manufacturers and their customers interacted through a process of risk assessment (21 per cent (14/66)), as discussed in Chapter 7.

Interactions between manufacturers and their customers contributed to firms' understanding of safety problems and ways to address them. Some customers also contributed to a general appreciation by manufacturers of the importance of machinery safety, through the customer's expectation that safety issues would be addressed. Manufacturers' methods for seeking and receiving input were rather basic compared with approaches to facilitating end user participation elaborated in the specialist body of knowledge, as outlined earlier in the chapter. Manufacturers made limited use of user trials or other methods for the purpose of anticipating unintended use and finding out how workers might struggle with the machinery. Also, when interaction between firms and their customers only occurred after the design stage, this interaction was not optimal for making more fundamental changes to the machinery and ensuring that it was inherently safe.

With regard to the construction of safety knowledge by those involved in machinery design and construction, variation in the nature of interactions between manufacturers and their customers engendered different sources of information and perspectives on safety problems and solutions. A firm that consulted and conducted user trials at the design stage, and involved end users and worker representatives, had information available to designers about the actual ways that end users interacted with the machinery and how it impacted upon them. The firm could use this information to design the machinery to be inherently safer. In contrast, a manufacturer that only received ad hoc feedback from managers in customer firms, after the machinery was in use in workplaces, only had information about whether the machinery functioned as intended. If this information included reports of injuries or incidents it might be used to redesign the machinery. Otherwise such feedback could only prompt the manufacturer to retrofit controls, or provide information to warn end users of hazards and the need to use safe work practices.

*Summary of Individual and Situational Differences*

Those involved in machinery design and construction learned about machinery safety matters through participation in these activities. In broad terms there were some common bases for learning; namely experience, interactions with customers, and technical standards. Due to differences in the professional and vocational backgrounds of key individuals and designers, and in design and construction practices in study firms, those involved in these activities constructed safety knowledge from multiple bases. The practices informing safety were the same as those applied more generally to address functionality, quality, cost, customer acceptance and other business goals. Practices such as inspection, consultation or prototype trials overlapped with methods required or encouraged in legal obligations (see Chapter 2), or advanced in the specialist body of knowledge for integrating safety in design (as above). However, as practiced in machinery design and construction, they were typically less developed and, as discussed below, only some practices in certain circumstances supported better performance for substantive safety outcomes.

### Practices and Individual Factors Linked with Better or Poorer Performance

To what extent then did the various bases for learning about machinery safety matters support good performance for hazard recognition, risk control and provision of safety information? The analysis in this section shows that some of the situational manifestations of practice (practices), and individual qualifications and experience (individual factors), were linked with better performance for these safety outcomes but others were linked with poorer performance.

The approach to analysis involved systematically reviewing the data about the performance of firms that engaged in particular practices, and the performance of firms with key individuals or designers with particular capacities.[1] This was the basis for reflecting on plausible relationships between levels of performance and particular practices or capacities, and inductively developing explanation which accounted for differences in performance for the substantive safety outcomes.

Of particular interest were the practices and individual factors linked with markedly better or poorer performance, when compared with the performance of firms in the sample overall. As demonstrated in Chapter 3, of the 66 study firms, 30 per cent had comprehensive hazard recognition; 14 per cent had a blinkered focus on mechanical hazards; 47 per cent used safe place controls as the primary risk control measures; 17 per cent used some advanced or innovative safe place controls; and 24 per cent provided substantial, good quality safety information. In order to distinguish more distinct trends, the focus here is on the practices and individual factors for which the proportion of firms performing at a particular level for a specific substantive safety outcome, was *at least 10 per cent above or below* the proportion of firms performing at that level in the sample overall.

Beginning with the key practice of interacting with customers, the findings are nuanced. While most firms interacted with their customers in some way, and a substantial proportion of firms actively sought input from their customers on safety matters, the only firms that performed better across a series of substantive safety outcomes were those that *sought input from end users at the design stage.*

---

1　The analysis was restricted to practices and individual factors for which there were ten or more firms that applied a particular practice, or in which particular individual factors applied.

As set out in Table 6.1 below, these firms were more likely to have comprehensive hazard recognition (63 per cent (10/16)), and to use safe place controls as the primary risk control measures (63 per cent (10/16)). They were also more likely to use some advanced or innovative controls (31 per cent (5/16)) and to provide substantial, good quality information (50 per cent (8/16)).

On the other hand the 25 firms that relied on customers to provide input, or received no input from customers or end users at all, performed more poorly compared with the sample overall. As set out in Table 6.1, these firms were less likely to have comprehensive hazard recognition (16 per cent (4/25)), and less likely to use safe place controls as the primary risk control measures (32 per cent (8/25)). Only one of these firms used more advanced or innovative controls (4 per cent (1/25)), and only one provided substantial, good quality information (4 per cent (1/25)).

The findings for customer/end user input suggest the potential for machinery manufacturers to perform better for substantive safety outcomes by actively seeking input from end users at the design stage, especially compared with simply relying on customers to provide input or receiving no input at all. The findings also suggest that interaction with end users, as practised by study firms, was not sufficient to ensure optimal safety outcomes. It is likely that manufacturers' rather basic methods for end user participation were a contributing factor. There was little evidence of the kind of practices advocated in the specialist body of knowledge as important for anticipating end user interactions with designed items, including unintended use, and for assisting end users to identify and articulate their concerns.

**Table 6.1    Interacting with customers and performance for substantive safety outcomes**

| | Substantive safety outcomes | | | | | | | | | |
|---|---|---|---|---|---|---|---|---|---|---|
| | Comprehensive | | Blinkered | | Safe place emphasis | | Advanced/ innovative | | Substantial, good info | |
| | n/N | % | n/N | % | n/N | % | n/N | % | n/N | % |
| Sought input from users at design stage | 10/16 | 63 | | | 10/16 | 63 | 5/16 | 31 | 8/16 | 50 |
| Whole sample | 20/66 | 30 | 9/66 | 14 | 31/66 | 47 | 11/66 | 17 | 16/66 | 24 |
| Received input, or no input | 4/25 | 16 | | | 8/25 | 32 | 1/25 | 4 | 1/25 | 4 |

*Note:* In Table 6.1, N is the number of firms using a particular practice, and n is the number of those firms performing at the stated level for the particular substantive safety outcome.

**Table 6.2     Other practices linked with better performance for several substantive safety outcomes**

| | Substantive safety outcomes | | | | | | | | |
|---|---|---|---|---|---|---|---|---|---|
| | Comprehensive | | Blinkered | | Safe place emphasis | | Advanced/ innovative | | Substantial, good info | |
| | n/N | % | n/N | % | n/N | % | n/N | % | n/N | % |
| Injury or incident info | 13/30 | 43 | 1/30 | 3 | | | 8/30 | 27 | 14/30 | 47 |
| Specialist or safety resources | 11/24 | 46 | 1/24 | 4 | | | | | 10/24 | 42 |
| Whole sample | 20/66 | 30 | 9/66 | 14 | 31/66 | 47 | 11/66 | 17 | 16/66 | 24 |

*Note:* In Table 6.2, N is the number of firms using a particular practice, and n is the number of those firms performing at the stated level for the particular substantive safety outcome.

Other practices linked with better performance across several substantive outcomes were *use of specialist or general safety resources,* and *use of information about injuries, incidents or hazardous situations.* As set out in Table 6.2 above, manufacturers that made use of specialist or safety resources were comparatively more likely to have comprehensive hazard recognition (46 per cent (11/24)), and less likely to be blinkered in approach (4 per cent (1/24)). They were also more likely to provide substantial, good quality information (42 per cent (10/24)). Manufacturers that used information about injuries, incidents or hazardous situations were more likely to have comprehensive hazard recognition (43 per cent (13/30)), and less likely to be blinkered (3 per cent (1/30)). They were also more likely to use some advanced or innovative control measures (27 per cent (8/30)), and more likely to provide substantial, good quality information (47 per cent (14/30)).

These findings suggest that use of specialist or general safety resources, and using information about injuries, incidents or hazardous situations helped to increase awareness in some firms of machinery hazards, or different instances of hazards, and of the need to provide safety information. Also, experience of injuries, incidents and hazardous situations may have encouraged some firms to contemplate more innovative or advanced control measures. The findings indicate, however, that the nature and scope of information available to firms through these sources was not optimal, and did not enable them to perform well across all substantive outcomes. It is likely that an unsystematic approach to seeking out authoritative sources of information about machinery safety and injury data, and ad hoc use of consultants, contributed to manufacturers' less than optimal performance.

Three other factors were linked with better performance but only for risk control outcomes. As set out in Table 6.3, manufacturers that *inspected the end use work*

**Table 6.3    Practices linked with better performance for risk control outcomes**

| | Substantive safety outcomes | | | | | | | | |
|---|---|---|---|---|---|---|---|---|---|
| | Comprehensive | | Blinkered | | Safe place emphasis | | Advanced/ innovative | | Substantial, good info | |
| | n/N | % | n/N | % | n/N | % | n/N | % | n/N | % |
| Supplier assistance | | | | | 13/18 | 72 | | | | |
| Inspection of end use workplace | | | | | 8/11 | 73 | | | | |
| Prototypes/ models | | | | | | | 9/33 | 27 | | |
| Whole sample | 20/66 | 30 | 9/66 | 14 | 31/66 | 47 | 11/66 | 17 | 16/66 | 24 |

*Note:* In Table 6.3, N is the number of firms using a particular practice, and n is the number of those firms performing at the stated level for the particular substantive safety outcome.

*environment* were more likely to use safe place controls as the primary risk control measures (73 per cent (8/11)), as were firms that *obtained or received assistance from component suppliers* (72 per cent (13/18)). Firms that *trialled prototypes or models* of their machinery, were somewhat more likely to use advanced or innovative control measures (27 per cent (9/33)). These findings suggest that these three practices may have offered insights about more effective risk control measures, or about the need for these.

The findings for professional and vocational background (individual factors) were mixed. They imply that engineers constructed more relevant safety knowledge than individuals with a trade background, or no qualification or trade, but that engineers' knowledge was also not optimal. As set out in Table 6.4, manufacturing firms with key individuals who were *engineers performed better on two substantive outcomes*. They were more likely to have comprehensive hazard recognition (50 per cent (11/22)), and to provide substantial, good quality information (36 per cent (8/22)).

On the other hand, as shown in Table 6.4, the firms with key individuals who were tradespeople, or who had no qualification or trade, were less likely to have comprehensive hazard recognition (18 per cent (6/34)), and less likely to provide substantial, good quality information (12 per cent (4/34)). Also, the firms with designers who were tradespeople, or who had no qualification or trade, were less likely to have comprehensive hazard recognition (20 per cent (4/20)), more likely to be blinkered (25 per cent (5/20)), less likely to use safe place controls as the primary risk control measures (35 per cent (7/20)), and less likely to provide substantial, good quality information (15 per cent (3/20)).

**Table 6.4        Individual factors linked with better or poorer performance for substantive safety outcomes**

| | Substantive safety outcomes | | | | | | | | | |
|---|---|---|---|---|---|---|---|---|---|---|
| | Comprehensive | | Blinkered | | Safe place emphasis | | Advanced/ innovative | | Substantial, good info | |
| | n/N | % | n/N | % | n/N | % | n/N | % | n/N | % |
| Engineer key individuals | 11/22 | 50 | | | | | | | 8/22 | 36 |
| Whole sample | 20/66 | 30 | 9/66 | 14 | 31/66 | 47 | 11/66 | 17 | 16/66 | 24 |
| Trade, or no qual. or trade key individual | 6/34 | 18 | | | | | | | 4/34 | 12 |
| Trade, or no qual. designer | 4/20 | 20 | 5/20 | 25 | 7/20 | 35 | | | 3/20 | 15 |

*Note:* In Table 6.4, N is the number of firms with key individuals or designers with particular qualifications (or not), and n is the number of those firms performing at the stated level for the particular substantive safety outcome.

The findings for professional and vocational qualifications are consistent with the notion that individuals with a trade background, or no qualifications or trade, had fewer skills for gathering, organizing, interpreting and applying information about machinery safety, and had a narrower or less relevant compendium of knowledge to draw upon to identify safety problems and determine what action to take. This notion is reflected in the comments of some tradesmen who referred to themselves as, "hands on people ... not intellectual people" (Managing director, Manufacturer 7) and as, "practical" people who "get out there and look after our own things" (Managing director, Manufacturer 20).

None of the other practices of machinery design and construction were linked with better or poorer performance for substantive safety outcomes. There was no evidence that visual inspection of machinery or computer aided drafting differentiated firm performance as, in each case, the performance of firms for particular outcomes was similar to the performance of the sample overall. Nor was there evidence that the central practice of referring to technical standards differentiated firm performance for substantive safety outcomes. As summarized in Table 6.5, the performance of these firms was similar to the sample overall with about one third comprehensively recognizing hazards (30 per cent (15/50)), some being blinkered (10 per cent (5/50)), half using safe place controls as the primary risk control measures (50 per cent (25/50)), a small proportion applying more

**Table 6.5    Referring to technical standards and performance for substantive safety outcomes**

| | Substantive safety outcomes | | | | | | | | |
| | Comprehensive | | Blinkered | | Safe place emphasis | | Advanced/ innovative | | Substantial, good info | |
| | n/N | % | n/N | % | n/N | % | n/N | % | n/N | % |
| Technical standards | 15/50 | 30 | 5/50 | 10 | 25/50 | 50 | [8/50] | [16] | 14/50 | 28 |
| Whole sample | 20/66 | 30 | 9/66 | 14 | 31/66 | 47 | 11/66 | 17 | 16/66 | 24 |

*Note:* In Table 6.5, N is the number of firms that referred to technical standards, and n is the number of those firms performing at the stated level for the particular substantive safety outcome.

advanced or innovative safe place controls (16 per cent (8/50)), and less than one third providing substantial, good quality information (28 per cent (14/50)).

There are two possible explanations for the finding that manufacturers that used technical standards did not achieve better performance for substantive outcomes. Those involved in design and construction activities typically referred to technical standards when they wanted specific information, rather than reviewing and applying standards systematically and rigorously. They may therefore have overlooked key standards, or key provisions in relevant standards. Also, technical standards may not adequately address all safety issues with machinery. European research has identified safety problems with automated installations arising from factors not addressed in relevant technical standards, as well as a series of weaknesses in the European harmonized standards for machinery (Backstrom and Döös, 2000; Boy and Limou, 2003, pp. 55–8, 88–9, 120; KAN, 2008). Analyses of Australian and international (ISO) standards for cranes and forklifts have also revealed weaknesses in their criteria for safe design and construction (Worringham, 2004; Lambert and Associates, 2003). It is likely that the ad hoc and piecemeal use of technical standards, coupled with specific weaknesses in particular standards, contributed to the unexceptional performance of firms that referred to technical standards.

Overall, the findings in this section indicate that the strongest links between better firm performance, and particular practices or individual factors, were for seeking input from end users at the design stage, using specialist or safety resources, using injury or incident information, receiving assistance from suppliers, inspecting the end use work environment, trialling prototypes or models, and having engineers as decision makers for machinery design and construction. These appeared to be sounder bases for constructing knowledge about machinery safety matters.

However, constructing knowledge about machinery safety from 'better bases' did not ensure that manufacturers performed well for substantive safety outcomes and therefore that they complied with the regulatory goal of prevention. Individuals might not have access to the kinds of experiences or sufficient experiences to construct comprehensive knowledge (Billett, 2008b). Even if a manufacturer engaged in practices generally linked with better performance for substantive outcomes, the firm might not perform well for particular outcomes if those making decisions and taking action on machinery safety matters did not work in or had not encountered situations that provided rich opportunities to learn about machinery safety. For example, even if designers consulted end users, what they learned might be minimal if the particular end users were unable to identify and articulate safety problems with the machinery they were consulted about.

The findings in this section also contribute to understanding whether non-state actors such as a firm's customers and suppliers positively influence the performance of the businesses with which they have commercial relationships. The wider safety and socio-legal literature suggests that parties in firms' supply chains and networks may help to build their capacity to comply with state regulation (Gunningham and Sinclair 2002, pp. 17–18; Hopkins and Hogan, 1998; Lamm and Walters, 2004; Walters, 2001, pp. 52, 375–6; Walters, 2002, pp. 45–6). The present research has shown that although the practice of interacting with customers and suppliers contributed to knowledge about machinery safety matters in some manufacturing firms, these interactions were only linked with better performance in specific circumstances. This was when suppliers contributed to capacity to identify and apply safe place risk control measures, and when manufacturers actively managed the interaction with their customers, rather than passively receiving information or directions. This research therefore reinforces studies indicating that the potential for non-state actors in firms' wider economic environments to exert a positive influence, and contribute to their capacity to comply with legal obligations, is circumscribed (Hutter, 2011, ch. 5; Hutter and Jones, 2007; James, et al., 2004).

Finally, differences in manufacturers' practices, and the professional and vocational backgrounds of key individuals and designers, also underlay variation in firm performance by size.[2] Most large firms had comprehensive hazard recognition (75 per cent (6/8)), and none were blinkered. Medium sized firms were less likely to have comprehensive hazard recognition (29 per cent (7/24)), as were small firms (21 per cent (7/34)). Most large firms used safe place controls as the primary risk control measures (88 per cent (7/8)), while medium firms were less likely to do so (42 per cent (10/24)), as were small firms (41 per cent (14/34)). The trend of better performance for substantive outcomes among large firms was also evident for provision of safety information, with most large firms providing substantial, good quality information (75 per cent (6/8)), while fewer medium sized firms (29 per cent (7/24)), and even fewer small firms did so (9 per cent (3/34)).

---

2    Small firms had less than 20 employees; medium firms had 20–99 employees; and large firms had 100 or more employees.

These differences were attributable to a mix of practices and individual factors which, taken together, meant that the capacity of the large firms to address machinery safety matters was stronger than the other firms. The large firms were more likely to use specialist or safety resources (63 per cent (5/8)), than the medium sized firms (46 per cent per cent (11/24)), and the small firms (24 per cent (8/34)). Large firms were also more likely to make use of information about injuries, incidents or hazardous situations (75 per cent (6/8)), than medium sized firms (46 per cent (11/24)), or small firms (38 per cent (13/34)). In addition, large firms were more likely to have key individuals who were engineers (88 per cent (7/8)), compared with medium sized firms (33 per cent (8/24)), and small firms (21 per cent (7/34)). No large firms had key individuals who were tradespeople or had no qualifications (0 per cent (0/8)), but half of the medium sized firms did (50 per cent (12/24)), as did many small firms (65 per cent (22/34)). Also, all large firms had engineer designers, typically working in large teams, but some medium sized firms had designers who were tradespeople or had no qualifications (29 per cent (7/24)), as did some small firms (38 per cent (13/34)).

Through differences in capacity, firm size shaped manufacturers' performance for substantive safety outcomes. The multi-faceted limits to resources and expertise in smaller firms are well recognized constraints on their ability to comply with safety regulation (Hutter, 2011, p. 147; Lamm and Walters, 2004; James, et al., 2004; Walters, 2001, p 32). In the present research, small and medium firms faced challenges in addressing machinery safety matters due to their more limited resources and expertise. They were less likely to have access to specialist or safety resources; less likely to engage in practices that supported learning about machinery safety matters; and their key individuals and designers were more likely to have vocational rather than engineering backgrounds, or no qualification or trade at all.

**Conclusion**

This chapter has demonstrated that there is a considerable body of specialist knowledge to support the integration of safety in the design and construction of machinery, but that study firms made little use of this. Human factors or other safety professionals may provide a conduit to facilitate access to this body of knowledge and facilitate its application, but without such support the practice of machinery design and construction was disconnected from the specialist knowledge base.

Rather than specialist sources, those involved in machinery design and construction principally learned about safety matters through participation in the activities of design and construction. In the course of these activities, they identified safety problems and determined what action to take in response to them along with functionality, quality, cost, customer acceptance of machinery and other business concerns. They drew upon their own and other firms' experience of producing machinery, interacted with customers and referred to technical standards, and constructed safety knowledge through these practices.

In addition, due to diversity in machinery design and construction practice, and the different personal histories and capacities of the individuals involved, learning and action on machinery safety matters differed between manufacturing firms. Some practices and individual factors were linked with better performance for hazard recognition, risk control and provision of safety information, but others were linked with poorer performance. Even where certain factors were linked with better performance they did not ensure optimal performance in all firms, or even in the majority of firms. Knowledge constructed through participation in design and construction practice, and on the basis of different professional and vocational histories, but disconnected from the specialist body of knowledge or other authoritative sources, was insufficient to support comprehensive hazard recognition, effective risk control, and provision of substantial, good quality safety information.

The findings in this chapter have significance for understanding the development of firms' capacity to self-regulate machinery safety functions and to comply with relevant legal obligations. They demonstrate that there are multiple bases from which individuals construct knowledge through practice, and for collective learning in firms. The constituents of knowledge go beyond and often outweigh specialist sources, as well as legal obligations for machinery safety, and information or advice from regulators (see also Chapters 4 and 5). These findings raise important implications for those seeking to build capacity for inherently safe design and construction in manufacturing firms.

There is a challenge for regulators and policy makers, as well as industry, professional and educational stakeholders, to contemplate how they might foster sound learning about safety matters through design and construction practice, rather than conceiving learning as based in individuals acquiring and reading regulatory instruments or guidance materials, participating in education and training in classroom settings, or seeking out and using specialist resources (see also Broberg, 1997; 2007). There is reason to suspect that the latter run the risk of being abstract representations of espoused practice that are remote from actual practice (see also Brown and Duguid, 1991). The issue of capacity building is explored further in Chapter 9.

There is also a need to consider the relationship between the practice of machinery design and construction generally, and specific practices for assessing and managing risks, as required for compliance with Australian and European legal obligations for machinery safety. This is the issue examined in the next chapter.

# Chapter 7
# Assessing and Managing Risks

To produce machinery that does not endanger health and safety, as required by the Australian and European regulatory regimes for machinery safety, manufacturers will need to institutionalize ongoing arrangements to systematically assess and manage risks in the course of machinery design and construction (see Chapter 2; see also Hutter, 2001, pp. 15–16, 77, 301–2; Johnstone and Jones, 2006). Such arrangements include processes for identifying hazards, assessing risks and determining risk control measures (risk management). For firms supplying machinery into Europe this entails assessing the conformity of their machinery with essential health and safety requirements and/or relevant harmonized standards, or arranging for a competent notified body to do this conformity assessment (European Commission, 1998a[1] art 8, cl 2, annex 1; DTI, 1999, pp. 4, 11; and see Chapter 2).

Processes for assessing and managing risks should therefore have been central to machinery manufacturers' practice. However, as we saw in earlier chapters, the relevant legal obligations and the activities of occupational health and safety (OHS) regulators were not key constituents of manufacturers' knowledge about machinery safety matters, and the regulators did not foster self-regulatory arrangements by manufacturers in their minimal inspection and enforcement with these parties (see Chapters 4 and 5). Also, those involved in machinery design and construction principally learned about safety matters, including risk assessment, through their interactions with customers and other industry contacts, and by referring to technical standards. (This is therefore also the reason that this chapter about assessing and managing risks follows the discussion of learning through practice in Chapter 6).

Was there, then, any evidence that manufacturers took proactive and systematic action to assess and manage risks in the course machinery design and construction? This is the issue addressed in this chapter. The insights provided are of relevance equally to readers interested in risk management and risk regulation generally, or in the specific areas of human factors, safety and design, and whether as specialists or practitioners, researchers, regulators or policy makers, educators or students in these fields.

An immediate challenge to the efficacy of risk management is the finding that the key decision makers for machinery design and construction in study firms (the key individuals), were not necessarily familiar with hazard, risk, assessment and

---

1 This directive was revised and reissued in 2006 for application in 2009, with essentially the same requirements for conformity assessment.

control concepts. Reflecting again their learning through practice, key individuals tended to use everyday expressions (safety problems, safety features), and some conflated hazard, risk and risk control (a see it/fix it approach). At the other end of the spectrum a small number of firms applied specialist FMEA and HAZOP[2] methods to systematically analyse accident scenarios and contributors. This divergence indicates the degree of segmentation in risk management capacity across study firms, from informal methods to complex risk analyses.

This chapter demonstrates the diversification in risk assessment practice, which arose from differences in the sources informing assessment, methods and timing, and the persons conducting assessment. Risk assessment could be done well or poorly, and was sometimes ritualistic (ticking boxes on a checklist). Better approaches supported logical, thorough and timely consideration of hazards and risk control measures. In particular, assessing risks at the design stage, and capturing different perspectives on risks by involving a group of people (within the firm, or among customers and end users), were practices linked with better performance across hazard recognition, risk control and safety information outcomes. However, risk assessment was often conducted too late to enable consideration of fundamentally different designs or integrated safeguards, sidelined to one or two design professionals or managers, or outsourced to consultants. Risk management was rarely part of a more systematic approach to managing safety along with other business risks and, regardless of the particular arrangements or methods used, performance for substantive outcomes fell short when those involved lacked the knowledge and skills for rigorous, well-informed decision making.

All of this is food for reflection about ways to build capacity for effectively assessing and managing risks in machinery design and construction. The chapter finishes by highlighting the practices, as identified in Chapter 6 and supplemented here, that were linked with better performance for substantive safety outcomes. These provide some ways forward for capacity building, which receives further consideration in the final chapter of the book.

### Risk Assessment and Risk Management Concepts

Technical approaches to risk management conceptually distinguish hazards, risks, assessment of risks and risk control measures (Cross, et al., 2000; Frick, 2006; and see CEN, 2010; NOHSC, 1994). Hazards are typically defined as the potential to cause harm (death, injury or illness), and risk as the combination of the probability (or likelihood) of the occurrence of harm and the severity of that harm (consequences). Assessment of risk therefore involves consideration of the severity of harm and the probability of their occurrence, but may be further broken down into a risk analysis phase involving hazard identification and risk

---

2  An FMEA is a fault mode and effects analysis and a HAZOP is hazard and operability study (Standards Australia, 1998).

estimation, and risk evaluation to determine measures needed to reduce the risk. Risk controls are the measures used to eliminate or minimize risks, ranging from design changes, engineering controls and safeguards, to warning signs or devices, safe work practices, and personal protective equipment (see Chapter 3).

Based on the accounts of the key individuals in the 66 study firms, only some were familiar with hazard, risk and risk assessment concepts. The key individuals in some firms did not use the term hazard (41 per cent (27/66)), and/or the term risk (26 per cent (17/66)). Also, these and some other firms did not conduct, or engage a consultant to conduct, any process which they called risk assessment or risks analysis (41 per cent (27/66)). When they did, their key individuals tended to use the terms assessment and analysis interchangeably, rather than seeing risk analysis as a phase within a broader process of risk assessment (as outlined above). It appeared that, in constructing knowledge about machinery safety matters through practice (see Chapter 6), key individuals had not worked in, or had experience of, situations through which they became familiar with hazard, risk and assessment concepts.

The absence of the terms hazard and risk from the discourse of some key individuals was more evident in firms that did not conduct any form of risk assessment but the key individuals in most firms used more everyday language. They talked about safety issues or safety problems (62 per cent (41/66)), and safety aspects or safety things (45 per cent (30/66)). Rather than discussing risk control measures, they referred to safety features, safety measures or safety devices (29 per cent (19/66)), or simply named particular types of measures such as guarding, covers, rails or safety switches.

The use of more everyday expressions by the key individuals in some firms was not just a matter of terminology. It reflected the conflation of concepts of hazard, risk and control measures in practice. As other research has found, individuals conducting assessments may lack the analytical abilities to separate cause and effect, and understand the principles of prevention (Jensen, 2001). In the present research there was a tendency to merge consideration of sources of harm with control measures in a see it/fix it approach. If a firm recognized the potential for someone to get hurt, they implemented preventive measures without considering, in a more differentiated way, the hazard, risk and possible risk control measures.

This undifferentiated approach was captured in the comments of a safety manager with a firm producing vehicle industry machinery. The firm had well-developed procedures for risk management, with differentiated process steps, but the firm's employees preferred a less structured approach. The safety manager stated:

> Just thinking about this hazard id, assessment and control process ... people don't understand it ... some are very thorough and you can tell they've gone through all the steps that they've been trained to do but it still doesn't mean a lot. What we do is just continual hazard elimination. Forget the id if you like, just keep looking, looking, looking and whenever you see a deficiency fix it and that's from the start of the process to the finish ... just continually eliminating

what we perceive will be the source of injury or accident and I think that's what people are doing, and the three steps seem to be imposed on top ... let people do what they do naturally and say, 'oh that could hurt something, we'll shave that off' or 'that could trip there, we'll get rid of that'. (Safety manager, Manufacturer 18)

The practice of managing risk through the lens of risk management, and the specific tool of risk assessment, are relatively recent developments in dealing with health and safety matters since the 1990s (Glendon, Clarke and McKenna, 2006, pp. 19, 414). It may simply take longer for risk assessment and risk management concepts to become part of the routine discourse and practice of firms. However, from a social constructivist perspective, learning occurs in communities of practice, and is characterized by facility with the discourse and practice associated with particular communities of practice (Lave and Wenger, 1990; Palincsar, 1998).

The findings about the conflation of key concepts of hazard, risk and control measures, and the endemic use of everyday expressions (safety issues, problems, aspects, things, features or measures) are evidence of the limited interaction between, on the one hand, OHS regulators and professional communities of practice and, on the other hand, the machinery manufacturer community of practice. As individuals in regulated firms were not part of the same interpretive community (Black, 1997, p. 30), they lacked an understanding of key concepts necessary to apply legal requirements effectively, and ensure machinery was safe and without risks to health. These findings reflect again the everyday practices and interactions through which individuals in manufacturing firms learned about machinery safety matters, and the fact that few firms engaged directly with relevant legal obligations, few firms had sought or received guidance materials or oral advice from OHS regulators, and the practice of machinery design and construction was largely disconnected from the specialist body of knowledge (see Chapters 4 to 6).

Risk assessment (or risk analysis) was part of the discourse and practice of machinery design and construction in a little more than half the study firms (59 per cent (39/66)). It was, however, ascribed different meanings and entailed different approaches as the 39 firms, or the consultants they engaged, adopted or adapted diverse risk assessment practices from different sources. The following section examines this diversity in the practice of risk assessment.

## Differences in the Practice of Risk Assessment

*Overview of Differences*

The diversification of risk assessment is well documented, both in safety practice and more generally (Frick, 2005; Harms-Ringdahl, 2001, pp. 43–54; Healey and Greaves, 2007; Horlick-Jones, 2005; Neathey, et al., 2006). This diversification was evident in the practices of the 39 manufacturers that conducted a form of risk assessment for their machinery, or engaged a consultant to do this. The practice of risk assessment varied as each firm's goals, interactions with external actors and other activities differed, and influenced how assessment took shape in the context of a particular firm's operations. In particular, diversity arose as manufacturers, or their consultants, adopted or adapted methods set out in different technical standards, imposed or provided by customers or other industry sources, required in the European regulatory regime for machinery safety or, more rarely, offered in guidance from Australian OHS regulators.

The analysis of risk assessment practice presented here is based on interviews with key individuals in the 39 firms that conducted some form of assessment, documentation of risk assessments which was available in 30 firms, and the key sources informing risk assessment practice. This analysis shows that as well as the various sources reflected in firms' risk assessments, the practice of assessment differed with regard to hazard identification and risk estimation methods, the relationship between risk assessment and risk control, the timing of assessment and who conducted the assessment. Some strengths of particular approaches are highlighted, as well as some inherent weaknesses of others.

*Sources Informing Assessment*

A variety of sources were reflected in the risk assessment practices of study firms and the consultants they engaged to conduct assessments. One was the Australian Standard, *Safeguarding of Machinery* (AS 4024, Standards Australia, 1996),[3] which was the basis of nine firms' risk assessments (23 per cent (9/39)). Seven used the standard themselves and two engaged consultants who applied the standard in their risk assessments.

The machinery standard AS 4024 provided a structured and detailed process for assessing and managing machinery risks. The process included a risk analysis phase in which assessors were to determine the limits of the machinery, systematically identify hazards with reference to information in the standard about different types of hazards, and estimate the level of risk based on information about the severity of harm and the probability of harm occurring, among other factors (Standards Australia, 1996, pp. 33–6). In determining the probability of harm, assessors were to consider the frequency and duration of exposure to the hazard, the probability of occurrence of a hazardous event, and the technical and human possibilities of avoiding or limiting harm. After estimating the level of risk, assessors were advised to evaluate risks to determine if risk reduction was required and, if so, to determine appropriate risk control methods. The standard provided advice about selecting different types of risk control measures according to the level of risk, and their effectiveness for eliminating and minimizing risks. It also provided some advice about producing safety information. If rigorously

---

3   Replaced by Standards Australia (2006) and now Standards Australia (2014).

applied, the machinery standard AS 4024 facilitated a logical and thorough process to assessing and managing risks.

A second source informing risk assessments was the Australian/New Zealand Standard 4360, *Risk Management* (Standards Australia, 1999).[4] Four firms' assessments were based on this standard (10 per cent (4/39)), which provided a generic guide to establishing and implementing risk management processes for addressing any business risks. It mapped out process steps of risk identification to determine how risks might arise, and risk analysis to determine possible consequences and likelihood in the context of existing controls, and to estimate the level of risk. The form of risk evaluation in AS/NZS 4360 involved comparing the estimated levels of risk against pre-established criteria, in order to rank or prioritize risks for risk treatment, including risk control. For firms applying this standard, a key element of assessment practice was the categorization of risk applying standardized descriptors of consequences (insignificant, minor, moderate, major, catastrophic), their likelihood (almost certain, likely, possible, unlikely, rare), and the level of risk (extreme, high, moderate and low) (Standards Australia, 1999, pp. 34–5). This approach to categorization of risk was prone to subjective, arbitrary and unreliable judgements by assessors, as discussed further below.

A third source from which two firms derived their risk assessment methods was guidance provided by OHS regulators (5 per cent (2/39)). One of these firms had been visited by an inspector during the South Australian OHS regulator's *Safer at the Source Intervention* and used a procedure based on guidance material provided by the inspector (see Chapter 5). The other firm had obtained the Victorian OHS regulator's plant hazard checklist and used this as the firm's risk assessment (WorkSafe Victoria, 2002). The checklist was a series of questions prompting users to identify hazards that might affect operators of machinery and other plant, anyone working in the vicinity, and others who could be affected. The checklist included the hazards of entanglement, crushing, cutting, stabbing, puncturing, striking, high pressure fluid, electricity, explosion, slipping, tripping and falling. It also covered ergonomic factors, suffocation, fire and thermal hazards, substances, noise, vibration and radiation, and prompted users to identify any other hazards not listed.

In addition to the manufacturer that obtained the plant hazard checklist from the Victorian OHS regulator, another seven firms (or their consultants) incorporated the same checklist of hazards in their risk assessments. They did so without identifying the original source but presented the same set of hazards, in the same sequence, in their risk assessment proformas. Through being passed on by consultants and other industry sources, the Victorian regulator's checklist was incorporated in the risk assessment processes of eight firms (21 per cent (8/39)).

A fourth approach to risk assessment was reflected in the practice of firms that supplied, or intended to supply, their machinery in Europe. This was conformity assessment with reference to the essential health and safety requirements in the

---

4   See Standards Australia (2004) and ISO (2009) for subsequent editions presenting a similar approach.

*Machinery Directive* and/or relevant European harmonized standards (European Commission, 1998a, annex 1; 2002). Nine firms conducted a conformity assessment for the European market, or engaged a consultant to do this (23 per cent (9/39)). Only three of the nine firms conducted the conformity assessment themselves and knew about the process required, the essential requirements or particular European harmonized standards for machinery. The other six firms engaged a consultant who specialized in the conformity assessment of machinery according to the European regulatory regime. These six firms handed the assessment over to the consultant and did not play an active part in the assessment process. With regard to harmonized standards, firms (or their consultants) applied European C type standards for particular types of machinery and/or the generic A type standard for machinery risk assessment (CEN, 2003a; 2003b).[5]

Apart from Australian Standards, the Victorian regulator's plant hazard checklist and the European approach to assessment, some study firms sourced methods from their customers or other industry contacts, or the consultants they engaged drew upon industry sources in fashioning their risk assessments. Wider customer or industry sources were reflected in the assessments of 15 firms (38 per cent (15/39)). Through these sources, firms or their consultants adopted diverse approaches to identifying hazards (one approach being the Victorian checklist), and for estimating risks. The latter embraced differing descriptors, numerical scales or rankings, and/or matrices, which were the basis for categorizing consequences (injury, harm or hazardous events), probability (likelihood) of these consequences occurring and the level of risk (see also 'Risk estimation' below).

For a further nine study firms, there was no evidence that they drew upon any sources to inform their risk assessment. They used an informal process of inspecting their machinery, described below, and simply applied the term risk assessment to their inspection (23 per cent (9/39)).

In summary, manufacturers and their consultants derived practices for risk assessment from technical standards, OHS regulator guidance materials (directly or indirectly through third party sources), the European *Machinery Directive* and harmonized standards, and industry contacts. They applied elements drawn from these different sources individually, or in combination. As a consequence, both the sources informing risk assessment and the practice of assessment varied between firms, as shown in the following sections.

*Informal Processes, Hazard Identification and Checklists*

For nine manufacturers, risk assessment was simply the term they used to refer to their informal process of inspecting machinery, thinking or talking through how people could get hurt, and deciding what they could do to reduce the possibility of harm (23 per cent (9/39)). The previous chapter showed that 33 firms inspected their own or similar machinery as a way to identify safety problems, solutions to

---

5    These were current at the time of data collection; now CEN (2010).

problems and opportunities to improve their machinery. Most of these firms did not call this inspection a risk assessment, but for nine firms the inspection was their risk assessment. As the manager of a firm that manufactured agricultural machinery stated, 'we look at the product, we assess the hazards mentally or physically and then address them before they go out ... but there's no formal writing'. (Manager, Manufacturer 33)

Other firms' risk assessments also included hazard identification. They variously identified hazards using a checklist of different types of machinery hazards (41 per cent (16/39)), and/or the European essential health and safety requirements or harmonized standards (23 per cent (9/39)), and/or they recorded the hazards identified for particular machinery (23 per cent (9 of 39)). Among the 16 firms that used a checklist, two simply ticked the boxes for relevant hazards on the checklist and added notations about the need for end users to use safe work practices or personal protective equipment. This constituted the risk assessment for these firms, which both used the Victorian OHS regulator's plant hazard checklist (WorkSafe Victoria, 2002). One firm had obtained the checklist from the regulator and the other obtained it from an industry contact. Both firms retained the completed checklist as a record of their assessment, and provided a copy to their customers and distributors. As the technical services manager of a firm that made construction machinery stated:

> ... that risk assessment [the checklist with boxes ticked] is supplied to all around
> Australia, to all our reps if they need a risk assessment ... and it's reassuring
> too for a purchasing officer ... If we send them a risk assessment we've done
> the right thing. They've asked for it, they've got it and they pass it on to the
> workshop foreman. (Technical services manager, Manufacturer 12)

This example captures the ritualistic nature of the tick-the-box approach to risk assessment. The purpose of such an assessment was for the manufacturer to be able to provide some paperwork to customers and distributors, rather than to comprehensively recognize hazards and determine effective preventive measures. Similar weaknesses were recognized in a Danish study of workplace assessment which found that organizations less successful with assessment generated paperwork based on checklists (Jensen, 2001).

Fourteen study firms included a hazard checklist together with additional elements in their assessment processes. Six of these 14 firms used checklists that included the same set of hazards as the Victorian regulator's checklist (WorkSafe Victoria, 2002), and in the same sequence, but without identifying the regulator as the source. Four firms used checklists which included a similar list of hazards but were based on the machinery standard AS 4024, three had developed their own checklists with the involvement of their safety personnel and one used a checklist developed by the South Australian OHS regulator.

In summary, machinery risk assessments included hazard identification in some form, and some assessments incorporated a checklist of hazards. While firms used different checklists, all examples provided a reference list of hazards, potentially prompting assessors to consider a wider range of hazards for the machinery.

*Risk Estimation*

Almost half of the manufacturers that conducted some form of risk assessment included an estimation of the level of risk (49 per cent (19/39)). This reflected a technical approach to risk assessment in which risk is estimated based on the probability and consequences of adverse outcomes, and then evaluated to determine the need for further preventive action (Brauer, 2006, pp. 645–53; Cross, et al., 2000; Frick, 2005; Glendon, Clarke and McKenna, 2006, pp. 17–18; Raafat, 1989). Six of these firms' approaches to risk estimation were based on the machinery standard AS 4024 (Standards Australia, 1996), two were based on the business risk management standard AS/NZS 4360 (Standards Australia, 1999), and two were based on a combination of these approaches. The remaining nine firms used alternative approaches, which they had adopted from customer or industry sources. These applied widely differing[6] descriptors, numerical scales or rankings, and matrices to categorize consequences, probability and level of risk. In some approaches assessors compared descriptors or numerical values for consequences and probability on a matrix, and identified the risk level for the corresponding consequence and probability categories. In others, assessors multiplied consequence and probability, or used another mathematical formula to estimate risk from consequence and probability values. The categorization approach in the customer and industry methods was prone to subjective, arbitrary and unreliable judgements, as with the approach in the business risk management standard AS/NZS 4360 (see 'Sources Informing Assessment' above).

The diversity of risk estimation approaches is illustrated by the comments of an engineer with a firm that manufactured packaging machinery. The firm had its own procedure for risk assessment based on the machinery standard AS 4024, but was also required by its customers to use their approaches, from time to time. The engineer stated:

> I know a few industries or companies use a risk assessment where you draw lines all over the page to match the different risks and then come up with a final rating and then it's up to you to decide. Then there's a list, if that rating is too high then you should do something about it. Then there's a couple where they

---

6   For example, alternative descriptors for consequences ranged from: insignificant/ negligible to catastrophic/fatalities; and for probability/likelihood alternatives ranged from rare/remote or highly/very unlikely/practically impossible to almost certain/common/ frequent/very likely. Alternative risk ranking systems included using a numerical scale to rank consequences, probability and/or level of risk as 1 to 5, 1 to 10, or otherwise. In some numerical approaches a high number signified a more serious consequence, or a higher probability or risk. In other approaches a high number signified a lower value for these variables.

use the building block idea ... Then there's others which use a whole heap of formulas and you just about have to be a mathematician or an engineer to go through it, and a lot of people find that's just too hard and they'll skip over that part of it. Even though it may end up a low risk ... you end up doing something about it anyway. So it would be good if there was some kind of standard ... .
(Engineer, Manufacturer 13)

As well as illustrating alternative approaches, this engineer's comments highlight the confusion engendered by multiple approaches to risk estimation, as well as the abstract nature of the estimation exercise. The character of such methods of risk estimation is at odds with decisions in legal proceeding under Australian OHS law which have highlighted the need for those conducting assessments to adopt an investigative and information seeking approach in order to determine the nature of hazards, how people could be exposed to them, and the available options for eliminating or minimizing risks (Bluff and Johnstone, 2005).

The subjective, arbitrary and unreliable nature of risk estimation is well documented in the safety literature which recognizes that anticipating future events is inexact, as data may not be available on consequences and probability, estimates of risk may be incomplete or based on past occurrences, and estimates depend on whose knowledge and perception of risk are applied (Cross, 2001; Frick, 2005; HSE, 2001b, pp. 28–9; Gadd, Keeley and Balmforth, 2003, pp. 26, 43). Conventional approaches to risk estimation have also been criticized for not dealing with factors other than the probability and consequences of adverse outcomes; factors such as the persistence of injury or damage, whether restoration is possible and long latency conditions (Klinke and Renn, 2002a; 2002b).

The Australian and European legal obligations for machinery safety did not require manufacturers to estimate risk by categorizing consequences, probability and level of risk. If anything they encouraged a more investigative approach, suggesting a series of factors to take into account in considering the severity and probability of harm, and using methods including inspections of machinery and the end use work environment, testing, analysis of injury and near miss data, and discussions with relevant parties (see Chapter 2).

Manufacturers' diverse approaches to risk estimation might give an illusion of a formal assessment of risk but did not necessarily support firms in achieving substantive safety outcomes. Their approaches were susceptible to subjective, arbitrary and unreliable judgements about consequences, probability and level of risk if assessors categorized these risk variables with reference to set descriptors, rankings or matrices. They also failed to recognize the limits to assessors' knowledge, did not rigorously seek information to strengthen their knowledge base, and failed to recognize that risk decisions inevitably vary depending upon whose knowledge and assumptions underpin them.

*Specialist Methods for Risk Analysis*

Four study firms applied specialist methods to systematically analyse what could go wrong with machinery based on identifying accident scenarios, accident contributors and ways to reduce risks before accidents happen (10 per cent (4/39)). Two of these firms used fault mode and effects analysis (FMEA) to analyse how each major part of their machinery could become faulty and the potential effects; and two used hazard and operability (HAZOP) studies to evaluate each component part, determine how deviations could occur and consider whether these could lead to hazards or operability problems (Reunanen, 1993, p. 10; Standards Australia, 1996, p. 153; 1998, pp. 21–4). The four firms that used FMEA or HAZOP studies all had large engineering design teams and produced more technologically sophisticated machinery or large-scale plant. An example was a firm that designed, commissioned and managed the construction of major plant installations. The firm's engineering manager explained:

> We also use, primarily more in the larger projects, but we do it in an informal manner for the smaller ones, a thing called a HAZOP … You have a special meeting and you can engage people that facilitate it … You visualize it or have scenarios and you say "if this fails what happens?" "Have we got protection on this and all those sort of issues?" So you do that schematically … and then you look at the physical arrangements and make sure you've got access issues covered and platforms and so on. (Engineering manager, Manufacturer 15)

As this example illustrates, specialist methods enabled those conducting the risk analysis to logically and systematically reflect on what could go wrong with machinery and the risks arising from failures. While such methods were not used by most study firms, some European research suggests that these techniques can be usefully applied in the design and redesign of machinery in order to identify additional accident scenarios, design solutions, safeguards, and information or warnings for end users (Reunanen, 1993, pp. 107–9; Swuste, et al., 1997). However, the application of specialist methods entails additional design resources, adds time to the design process and requires the support of experienced personnel, which are constraints likely to limit the application of these techniques in machinery design and construction, particularly in firms lacking engineering design teams with risk analysis expertise.

*Criteria for Determining Risk Control Measures*

Determining risk control measures was an element of all study firms' risk assessments except for the two firms, discussed above, that simply ticked off hazards on a checklist. If the risk assessment was recorded in some form, the risk control measures were also set out in the assessment report. Only two firms referenced the reasonably practicable standard from Australian OHS law as the criterion for determining the level of risk control (5 per cent (2/39)). That so few firms took account of this legal criterion is unsurprising in view of the generally low awareness of the legal obligations for machinery safety among key individuals in study firms (see Chapter 4).

Nine firms (23 per cent (9/39)) applied criteria provided in the machinery standard AS 4024 in determining risk control measures. For example, a firm that manufactured vehicle component production machinery had developed a procedure, which explained how the firm's design engineers should check and select risk control measures based on risk categories in this standard. An engineer with the firm explained:

> We have our own check sheet ... this refers to categories and tables which are in the document [AS 4024], so that if somebody picks a particular risk category then already it's explained in here ... before they start that checklist the first thing they do is they go to this simple process of using the standard and that shows how we've selected this category. So we've copied this onto a computer screen so that when somebody opens the file they can go to the back page [of AS 4024] and see how we went about getting, in just basic terms, just going through these three severity lists to get that [type of control]. (Engineer, Manufacturer 11)

If rigorously applied, the machinery standard AS 4024 provided a sound technical basis for determining risk control measures. It encouraged elimination of hazards in the design, and provided detailed advice about control systems and devices to minimize risks, and the selection of controls proportionate to risk (Standards Australia, 1996, chs. 6–10). The approach was compatible with the Australian legal criterion to control risks so far as reasonably practicable.

One manufacturer's assessment methods incorporated the firm's own criteria for evaluating whether preventive action should be taken, and the type of action, based on the estimated level of risk (3 per cent (1/39)). The firm's approach called for risks categorized as level 1 to 3 to be eliminated, risks categorized as level 4 or 5 to be eliminated if practicable, and risks categorized as level 6 to be eliminated only if such measures were cost neutral. This example of alternative criteria for evaluating risk, and determining preventive measures, again illustrates the arbitrary nature of some risk decision making, as well as inconsistency with the principle of eliminating or minimizing risk so far as reasonably practicable.

In addition to the legal, technical or alternative criteria applied by some manufacturers in determining risk control measures, all of the firms that conducted risk assessment, as well as those that did not, made choices about control measures which were shaped by their economic motivations and the attitudes of key individuals in firms. They determined preventive measures that were known, available, and that they considered compatible with machinery functionality, cost, competiveness or other business goals, and the attitudes of key individuals about the firm's responsibility (or not) for risk control. This issue receives further attention in Chapter 8, which examines the motivational factors that shaped firms' responses on machinery safety matters, and their choice of risk control measures.

*The Timing of Risk Assessment*

The specialist body of knowledge emphasizes the importance of addressing safety matters from the earliest stages of design in order to enable fundamentally different designs which eliminate hazards, and to ensure that risk control measures are compatible with the design, not readily removable by end users, and not likely to be removed or disarmed because they are a hindrance (Kletz, 1998b; Polet, Vanderhaegen and Amalberti, 2003; Reunanen, 1993, p. 108; Sagot, Gouin and Gomes, 2003; Swuste, van Drimmelen and Burdorf, 1997). The Australian and European regulatory regime for machinery safety also required that machinery was designed and constructed to be safe and without risks to health (not endanger health and safety), in the first instance (see Chapter 2). There were therefore legal as well as practical imperatives to address safety matters early in the life cycle of machinery.

While 15 study firms began assessing risk at the design stage (38 per cent (15/39)), and two firms conducted assessment from the construction stage (5 per cent (2/39)), there was a strong tendency for firms to leave assessment until a later stage. Fourteen firms did not assess risks until the supply or installation of machinery (36 per cent (14/39)), and eight firms did not assess risks until after the machinery was in use in workplaces (21 per cent (8/39)).

It is important to emphasize that failure to conduct risk assessment at the design or construction stages did not mean that a manufacturer took no action to address machinery safety in these early life cycle stages. Those involved in design and construction might review safety issues with reference to diagrams or plans during computer aided design, when trialling prototypes or models, or by other means (see Chapter 6). However, the activity that firms called risk assessment was distinct from these activities and was often done too late to enable fundamentally different designs, or to ensure that safeguards were integral to and compatible with the design and function of the machinery. Assessments conducted as late as supply, installation or end use might at best prompt manufacturers to retrofit risk control measures or bring safety issues to their attention, which they might address at a later date if they developed new models of their machinery.

*Persons Conducting Risk Assessment*

In the OHS field it is generally recommended that teams of people should conduct risk assessments in order to gather the collective knowledge, skills and experience of a range of people, and to take account of different interests, perceptions and assumptions about risks (Gadd, Keeley and Balmforth, 2003, p. 19; Glendon, Clarke and McKenna, 2006, pp. 34–5; Walters and Frick, 2000). Team-based assessment, together with rigorous gathering of information about hazards and risk control measures, may be a firm's best protection against arbitrary, subjective and unreliable risk assessment. Exclusively relying on consultants to conduct risk assessment is also a concern as firms that outsource assessment may not learn about the process themselves, and because consultants may not have adequate knowledge of the system assessed (Gadd, Keeley and Balmforth, 2003, p. 20).

With regard to who conducted machinery risk assessment, the practices of most study firms were weak in several ways. Although 31 firms conducted assessments in-house (79 per cent (31/39)), 25 of these firms involved only one or two people in their assessments. Those involved were usually design or project engineers or draftsmen, managers responsible for product development or production, or other managers. Only six firms used a team of people to conduct risk assessments (15 per cent (6/39)), involving engineers, production and/or safety personnel in their assessments (but see below, some firms involved their customers or end users).

A further seven firms engaged consultants to conduct all of their machinery risk assessments (18 per cent (7/39)), and one firm engaged a consultant to conduct some assessments (3 per cent (1/39)). The firms that engaged consultants did not involve their own employees in the assessment process, and the consultants did not have full knowledge of potential safety problems with the machinery as they did not, with one exception, consult the firms' customers or end users of the machinery. The key reason that manufacturers chose to use consultants was because they perceived that risk assessment required particular expertise which was not available in their firms, and they did not believe that it was worthwhile or that they were capable of developing the necessary expertise in-house. For example, a firm that supplied its machinery within Australia and overseas, including to Europe, conducted risk assessment about once each year when the firm produced a new item of machinery. The firm's engineering manager and engineer respectively stated that:

> [The consultant] is an expert at it so he was contracted to do a formal risk assessment of the product. (Engineering manager, Manufacturer 55)

> I would like to continue to use an expert on it ... I think it's necessary for us, rather than creating experts within the building. I mean this process happens once a year. (Engineer, Manufacturer 55)

A specific area of expertise for which manufacturers engaged consultants was conformity assessment for the European market, as required by the European regulatory regime for machinery safety. Of the eight firms that engaged a consultant to conduct all or some risk assessments, four used a consultant to assess the conformity of their machinery to European requirements, and two others used a consultant for assessment under European as well as Australian OHS law.

The number of firms that sought input to the risk assessment process from their customers or end users of their machinery was small. Nine firms involved their customers in their risk assessments (23 per cent (9/39)), and six of these involved actual end users (15 per cent (6/39)). There were separate instances of manufacturers' personnel being involved in risk assessments initiated by their customers (13 per cent (5/39)). Overall, customer or end user involvement in risk assessment was less common than some other forms of customer input such as participation in user trials, consultation and ad hoc feedback from people with experience of machinery in end use (see Chapter 6).

For most of the firms that conducted risk assessment, or engaged a consultant to do this for them, the practice of risk assessment was 'in the side-car', a term coined by Nordic researchers to highlight the separate treatment of safety functions, rather than integrating them in business management (Jensen, 2002b). In study firms risk assessment was typically sidelined in the activities of one or two people in the firm or outsourced, and was not routinely integrated in the interactions between manufacturers and their customers.

## Some Observations about Risk Assessment Practices

There was considerable diversity in the meaning and practice of risk assessment as developed, adopted or purchased by study firms, and drawing upon particular technical standards, customer and other industry sources, the European regulatory regime for machinery safety or, more rarely, Australian OHS regulators. Some risk assessment practices had the potential to support the logical, thorough and timely consideration of hazards, risks and control measures by firms. These practices included: identification of hazards using comprehensive checklists of machinery hazards; more structured and information seeking approaches, such as that presented in the machinery standard AS 4024; assessment at the design stage; and assessment involving a range of people. Intuitively, these practices might be expected to strengthen firms' decision making and action on machinery safety matters, and their performance for hazard recognition, risk control and provision of safety information.

Other approaches to risk assessment potentially narrowed the focus of assessment or increased the possibility of manufacturers making arbitrary, unreliable or subjective decisions. These practices included: informal assessments; a pre-occupation with categorizing risk variables according to risk descriptors, rankings and/or matrices; the practice of using only one or two people to conduct assessments; lack of involvement of customers or end users; outsourcing assessment to external consultants without any in-house involvement; and delaying assessment until supply, installation or end use of machinery.

Some of the different features of risk assessments were linked with firms achieving better performance for the substantive safety outcomes, and others were not, as demonstrated later in this chapter. First, the next section examines the extent to which risk assessment was part of a wider systematic approach to managing machinery safety matters in study firms.

## Systematic Management of Safety in Machinery Design and Construction

The safety and self-regulation literature more generally suggests that systematic management of business functions may facilitate organizations' compliance with safety, and other social and business regulation (Frick and Wren, 2000; Parker, 2002, pp. ix–x, 43–61; Parker and Gilad, 2011). In the field of safety, systematic management involves organizations accepting responsibility for safety, building capacity, and ensuring sufficient human and financial resources in order to implement, maintain, monitor and review organizational arrangements for managing safety, including risk management and worker participation (Hale, 2003; Johnstone, Bluff and Clayton, 2012, pp. 28–31; Nytrö, Saksvik and Torvatn, 1998; Saksvik and Quinlan, 2003; Zwetsloot, 2000).

There are various national and international standards or guidelines setting out arrangements for systematic safety management (BSI, 2007; 2008; CSA, 2006; ILO, 2001; SAA, 1997; 2001b; Walters and Jensen, 2000). In Australia workers' compensation authorities have encouraged larger organizations to implement safety management arrangements as a means to reduce workers' compensation claims and levies or, in a form of meta-regulation, required organizations to implement safety management standards as a condition for self-insurance (see for example Victorian WorkCover Authority, 2002b; WorkCover Corporation, 2001a; 2001b; 2008; 2013; WorkSafe Victoria, 2008; and see Gunningham, 2010, p. 135).

The various safety management standards and guidelines have largely focused on employers managing safety for their workforces, and have generally paid little or no attention to the management of safety in the upstream functions of design, manufacture, import and supply (Bluff, 2001). However, early UK guidelines encouraged organizations to manage safety in the manufacture and supply of products (HSE, 1997, p. 36) and, more recently, the American National Standards Institute (ANSI) in the US has issued guidelines for addressing safety in design and redesign processes (ASSE, 2011).

Analysis of interviews, documented procedures and arrangements for the 66 study firms revealed that four firms, all large, managed safety functions on an ongoing basis. Three of these firms were self-insurers under a state workers' compensation scheme and, in order to maintain their self-insurer status, implemented the scheme's safety management standards (WorkCover Corporation, 2001a, pp. 9, 15–16, annexure B). The fourth firm had implemented arrangements for managing safety applying the Australian Standard for OHS management systems, AS 4804 (SAA, 1997).

These four firms integrated the management of safety, quality and business risks more generally. They addressed the safety aspects of machinery design and construction in their safety management arrangements, although this was not required by the workers' compensation standards, and the relevant Australian Standard (AS 4804) only briefly advised suppliers of goods to address safety at each stage of the design cycle, and to have risk management procedures that controlled risks to customers (SAA 1997, pp. 19, 25–6; 2001, p. 10).

In each of the four firms that managed safety more systematically the key individuals emphasized senior management's commitment to safety matters. For example, key individuals in two firms respectively stated:

> ... safety is driven from the top down and as you can see on our values safety is always mentioned ... people who don't take safety on as part of the nature of the way they go about their work day to day then they don't really survive in this company. (Engineering manager, Manufacturer 11)

> I've never worked anywhere or heard of anywhere that is quite as committed to the safety function as what I have here. And I do have direct entrance and access to the MD ... So safety is not seen as one of these things as forced from outside. It's one of the factors that are in-built into the culture of at least the management sector. So when we developed the vision and the underlying goals and targets there, right at the top of that list was that we'll make sure that our processes are as safe as we can make them. (Safety manager, Manufacturer 15)

These examples are typical of the way that senior management accepted responsibility for and drove the management of safety in these four firms. There was also evidence that each of the four firms took steps to build capacity for addressing safety in machinery design and construction. All four firms had in-house safety managers whose role encompassed the safety aspects of machinery design and construction. For their staff involved in design and construction activities, three of the firms provided access to specialist resources, and all four provided safety training and access to comprehensive collections of technical standards. As the business development manager in one firm explained:

> ... our guys by and large are highly trained. We run a lot of in-house workshops on bringing people up to speed on various aspects of the safety. ... We do a lot of training sessions with people here, our new designers, and we certainly put them through their paces to make sure they understand fully. (Business development manager, Manufacturer 56)

Developing safety capacity, including the capacity of those involved in machinery design and construction, was a priority in the four firms that managed safety systematically. These firms also had a series of arrangements in place for the ongoing management of safety functions, including decision making and action at the design and construction stages for machinery. Each firm had documented procedures for risk assessment that involved assessment during design and construction. They practised risk assessment on an ongoing basis, each time they produced new machinery, and recorded their assessments. Each firm collected and analysed information about injuries and other incidents involving their type of machinery, and had arrangements for testing or checking the safety aspects of machinery, and for producing machinery safety information. The four firms also managed inputs to machinery design and construction from suppliers or contractors, including the safety aspects of these, and three of the four firms had procedures for seeking input from end users during machinery design.

It is not surprising that only four firms managed safety systematically and that they were all large firms. The safety and socio-legal literature indicates that enterprise size is a factor in the implementation of safety management and other self-regulatory compliance programmes, as such programmes are typically developed by or for larger organizations and unlikely to be suitable for smaller ones (Hale, 2003; Hale and Hovden, 1998; Parker, 2002, p. 56; Zwetsloot, 2000).

European research into safety in smaller enterprises has identified a series of firm characteristics that may hinder their management of safety functions (Walters 2001, pp. 32–52). In particular, managers in smaller firms tend to have too much to handle, concentrate on production matters immediately in front of them, and have difficulty allocating human and financial resources to areas other than production. They also tend to rely on informal rather than formal structures and processes, and are not inclined to seek external advice from consultants or advisers, unless they first establish trust in the source. All of these factors are likely to militate against the development of commitment, capacity and arrangements to manage safety functions systematically in smaller enterprises. On the other hand smaller firms are capable of implementing simplified approaches, which focus on the development of safety knowledge and skills, participation by and communication with workers, hazard identification and the development of control options, and action plans to track implementation (Hale and Hovden, 1998; Lamm and Walters, 2004; Pearse, 2001; Walker and Tait, 2004).

The present research has suggested that similar types of simplified practices, already applied in machinery design and construction, are feasible for smaller firms and are linked with achieving better performance for the substantive safety outcomes for hazard recognition, risk control and provision of safety information. As shown in Chapter 6, in addition to having an in-house person trained in safety matters, these feasible practices include: seeking input from end users at the design stage; collecting and using injury, incident and hazardous exposure information provided through customer feedback; seeking information from suppliers of safety components about state-of-the-art risk control measures; inspecting the end use work environment; and trialling prototypes or models, especially with end users. Consistent implementation of these practices by manufacturing firms may provide an alternative basis for institutionalizing proactive and systematic arrangements to manage safety in design and construction functions, which are feasible for a cross-section of firms.

The following section shows that certain risk assessment practices also supported better performance by manufacturers for the substantive safety outcomes. These practices may be usefully integrated with the other feasible practices as arrangements for addressing safety in machinery design and construction on an ongoing basis.

**Assessment Practices Linked with Better or Poorer Performance**

So far this chapter has demonstrated that a little more than half the study firms conducted a process of risk assessment, or engaged a consultant to do this for them, and firms and their consultants applied diverse approaches to risk assessment. Few firms embedded risk assessment in more developed systems for managing safety matters on an ongoing basis but, as set out below, certain practices for assessing risks were linked with markedly better or poorer performance for hazard recognition, risk control and safety information outcomes.

The approach to analysis involved systematically reviewing the data about the performance of firms that engaged in particular assessment practices,[7] as the basis for reflecting on plausible relationships between levels of performance and particular practices or capacities, and inductively developing explanation which accounted for differences in performance for the substantive safety outcomes. Of particular interest were assessment practices linked with markedly better or poorer performance, when compared with the performance of firms in the sample overall. As demonstrated in Chapter 3, of the 66 study firms 30 per cent had comprehensive hazard recognition; 14 per cent had a blinkered focus on mechanical hazards; 47 per cent used safe place controls as the primary risk control measures; 17 per cent used some advanced or innovative safe place controls; and 24 per cent provided substantial, good quality safety information. In order to distinguish more distinct trends, the focus here is on assessment practices for which the proportion of firms performing at a particular level, for a specific substantive safety outcome, was *at least 10 per cent above or below* the proportion of firms performing at that level in the sample overall.

For the 39 firms that conducted risk assessment, or engaged a consultant to do this, there was little evidence that assessment in itself differentiated firms' performance for substantive safety outcomes. Compared with the sample overall the performance of these 39 firms was only better for the outcome of hazard recognition, with more of these firms having comprehensive hazard recognition (41 per cent (16/39)), as set out in Table 7.1 below. The performance of the risk assessing firms was not markedly different from the sample overall for the other outcomes.

There was also some evidence to support the notion, introduced above, that *checklists* might prompt assessors to consider a wider range of hazards for machinery. As shown in Table 7.1, the firms that identified hazards with reference to a checklist of machinery hazards were more likely to have comprehensive hazard recognition (44 per cent (7/16)), and none of these firms were blinkered (0 per cent (0/16)).

The best performance across all substantive safety outcomes was achieved for assessments *conducted by a group of people*, either by a team within a

---

7    The analysis was restricted to practices for which there were ten or more firms that applied the particular practice.

**Table 7.1**     **Risk assessment practices linked with better performance for hazard recognition only**

| | Substantive safety outcomes | | | | | | | | | |
|---|---|---|---|---|---|---|---|---|---|---|
| | Comprehensive | | Blinkered | | Safe place emphasis | | Advanced/ innovative | | Substantial, good info | |
| | n/N | % | n/N | % | n/N | % | n/N | % | n/N | % |
| Hazard checklist | 7/16 | 44 | 0/16 | 0 | | | | | | |
| Any form of risk assessment | 16/39 | 41 | | | | | | | | |
| Whole sample | 20/66 | 30 | 9/66 | 14 | 31/66 | 47 | 11/66 | 17 | 16/66 | 24 |

*Note:* In Table 7.1, N is the number of firms adopting an assessment practice, and n is the number of those firms performing at the stated level for the particular substantive safety outcome.

manufacturing firm, and/or by a firm involving customers or end users in the assessment process. As set out in Table 7.2 below, firms whose assessments were conducted by a group were more likely to have comprehensive hazard recognition (40 per cent (4/10)), and none were blinkered (0 per cent (0/10)). They were also more likely to use safe place controls as the primary risk control measures (70 per cent (7/10)), to use advanced or innovative control measures (30 per cent (3/10)), and to provide substantial, good quality safety information (60 per cent (6/10)).

There was also some evidence to suggest that manufacturers whose assessments were *conducted at the design stage* performed better for a mix of outcomes. As shown in Table 7.2, these firms were more likely to have comprehensive hazard recognition (47 per cent (7/15)), to use safe place controls as the primary risk control measures (60 per cent (9/15)), and to provide substantial, good quality information (40 per cent (6/15)). However, these firms were not less likely to be blinkered in their hazard recognition, and nor were they more likely to use advanced or innovative control measures (13 per cent (2/15)).

Firms whose assessments were more *structured*, including specific hazard identification and risk estimation steps, performed better for hazard recognition and provision of safety information. As also set out in Table 7.2, more of these firms had comprehensive hazard recognition (53 per cent (10/19)), none were blinkered (0 per cent (0/ 19)), and more of these firms provided substantial, good quality safety information (58 per cent (11/19)). However, they did not perform optimally for hazard recognition and safety information, and for risk control outcomes their performance was not markedly different from the sample overall. The emphasis in structured assessments on categorizing risk variables using

**Table 7.2    Risk assessment practices linked with better performance for several substantive safety outcomes**

| | Substantive safety outcomes | | | | | | | | | |
|---|---|---|---|---|---|---|---|---|---|---|
| | Comprehensive | | Blinkered | | Safe place emphasis | | Advanced/ innovative | | Substantial, good info | |
| | n/N | % | n/N | % | n/N | % | n/N | % | n/N | % |
| Group assessment | 4/10 | 40 | 0/10 | 0 | 7/10 | 70 | 3/10 | 30 | 6/10 | 60 |
| Design stage | 7/15 | 47 | | | 9/15 | 60 | | | 6/15 | 40 |
| Structured assessment | 10/19 | 53 | 0/19 | 0 | | | | | 11/19 | 58 |
| Whole sample | 20/66 | 30 | 9/66 | 14 | 31/66 | 47 | 11/66 | 17 | 16/66 | 24 |

*Note:* In Table 7.2, N is the number of firms adopting a particular assessment practice, and n is the number of those firms performing at the stated level for the particular substantive safety outcome.

descriptors, rankings and/or matrices, rather than adopting an information seeking approach to determine hazard exposures and risk control options may explain, at least in part, the less than optimal performance of the firms that included such risk estimation methods in their assessments.

On the other hand, the firms that adopted *informal approaches* to risk assessment performed most poorly across a series of substantive safety outcomes. These informal methods were risk assessment in name only as they involved inspection of machinery or ticking off hazards on a hazard checklist (a tick-the-box approach). As set out Table 7.3, these firms were less likely to have comprehensive hazard recognition (9 per cent (1/11)), less likely to use safe place controls as the primary risk control measures (36 per cent (4/11)) and less likely to use advanced or innovative controls (9 per cent (1/11)). None of these firms provided substantial, good quality safety information (0 per cent (0/11)).

Considering the findings for risk assessment in aggregate, they suggest that it may be helpful to integrate the practices of conducting assessment at the design stage, involving groups of people and using checklists to prompt attention to a wider range of hazards, in order to sustain better performance across all substantive safety outcomes. These methods complement the feasible ways to institutionalize attention to safety in machinery design and construction, by small, medium and large firms, as summarized in the previous section (and identified in Chapter 6).

It is important to emphasize, however, that risk assessment practices linked with better performance for substantive safety outcomes did not ensure optimal performance. Nor did the firms that used informal approaches all have poor performance for each safety outcome. Risk assessment was neither a panacea nor

*Safe Design and Construction of Machinery*

**Table 7.3    Informal approaches linked with poorer performance for substantive safety outcomes**

| | Substantive safety outcomes | | | | | | | | | |
|---|---|---|---|---|---|---|---|---|---|---|
| | Comprehensive | | Blinkered | | Safe place emphasis | | Advanced/ innovative | | Substantial, good info | |
| | n/N | % | n/N | % | n/N | % | n/N | % | n/N | % |
| Whole sample | 20/66 | 30 | 9/66 | 14 | 31/66 | 47 | 11/66 | 17 | 16/66 | 24 |
| Informal (inspection or tick-the-box) | 1/11 | 9 | | | 4/11 | 36 | 1/11 | 9 | 0/11 | 0 |

*Note:* In Table 7.3, N is the number of firms adopting informal assessment practices, and n is the number of those firms performing at the stated level for the particular substantive safety outcome

a universal problem. The practice of risk assessment differed between firms but so too did the knowledge base from which those involved in machinery design and construction, and their consultants, conducted assessments. As the safety manager of a firm that manufactured vehicle industry production machinery stated:

> ... the risk assessment is a tool that we can teach engineers how to use, and that's what it is, a tool. But to make that tool work effectively you need to have the ability and the experience to identify potential hazards, and sometimes engineers don't always have that ability ... so that's knowledge and experience, and then you've got the risk assessment process and then out of that you've got identifying appropriate control measures, which again is experience and knowledge based ... we can help them out with the thing in the middle, the actual tool, but the two things on either end – it's not something you can teach people, that's my view anyway ... The quality of the risk assessment comes back to the individual and actually using everything that's available to them. (Safety manager, Manufacturer 18)

This manager's comments highlight the contribution of the knowledge base of individuals conducting risk assessment. Differences in this knowledge base, as demonstrated in Chapter 6, contributed to variation in performance for the substantive safety outcomes.

With regard to more developed systems and arrangements for managing safety, as there were only four firms with such arrangements it was not possible to determine whether systematic management differentiated firm performance for substantive safety outcomes. However, the mixed performance of the four firms for these safety outcomes is noteworthy. All four firms provided substantial, good quality safety information, but only two had comprehensive hazard recognition and used safe place controls as the primary risk control measures. The other two firms either had incomplete hazard recognition, or relied exclusively on safe person measures for some risks. Like risk assessment, systematic safety management was not a panacea guaranteeing good performance for substantive safety outcomes.

In essence, preventive action by manufacturers, including the assessment of risks and management of safety matters, fell short of what was needed to perform well for all machinery safety outcomes, and substantively comply with the regulatory goal of preventing death, injury and illness, when those involved lacked the knowledge to support well-informed and rigorous management of risks. Further, as demonstrated in the next chapter, firms' responses on machinery safety matters were also the product of motivational factors, including the legal, quasi-legal and economic motivations of firms, and the attitudes of key individuals, which influenced firms' decision making and actions. Risk assessment was only a key method contributing to firms performing well for substantive safety outcomes when it involved a structured assessment conducted at the design stage by teams of people and involving end users, when these practices were underpinned by a sound knowledge base for addressing machinery safety matters, and motivational factors provided the space for assessors to choose effective preventive measures.

## Conclusion

There is cause for concern that some of the study firms did not conduct any kind of risk assessment. There is also reason to suspect that more firms conducting risk assessments would not, in and of itself, achieve the goal of inherently safe design and construction of machinery. Differences in the sources informing risk assessments, methods used, timing of assessment and the persons conducting them all had implications for the quality of assessment. As a consequence risk assessment was a process that could be done well, in a mediocre fashion or quite poorly.

Safety scholars have suggested that risk assessment may be a key method in making self-regulation work and contributing to a better work environment, or a paper tiger that contributes to the bureaucratization and technocratization of safety policy and management (Karageorgiou, et al., 2000). The present research provides empirical evidence supporting both of these hypotheses.

At its best, risk assessment assisted manufacturers to comprehensively recognize hazards and determine preventive measures. The process reflected an informed, logical and structured way to systematically identify and examine hazards, determine the adequacy of existing controls and whether additional preventive measures were needed, and to compare different risk control options (see also Gadd, Keeley and Balmforth, 2003, p. 6; Raafat and Sadhra, 1999; Standards Australia, 1996). As such, risk assessment could contribute to firms performing well for substantive safety outcomes.

At its worst, risk assessment was little more than a tick and flick exercise in which the assessor ticked boxes on a checklist to identify hazards (not necessarily all of them), added notations to the checklist advising end users to use safe work practices or personal protective equipment, and passed the so-called risk assessment on to customers and distributors. This ritualistic response was risk assessment as a paper tiger. When firms' motivations for conducting assessment did not align with the goal of designing and constructing safe machinery, assessment was perfunctory and symbolic rather than substantive (see also Hutter, 2005; Jensen, 2002b).

Between the best and the worst examples of risk assessments there were a number of ways in which assessment was inadequate. It was often conducted by one or two people in a firm, not integrated with other design and construction activities, and in the absence of proactive, systematic and ongoing action to manage machinery safety matters. Assessment was sometimes not conducted until machinery was supplied or installed, or even after it was in use in workplaces. Some firms relied on consultants to conduct assessments, rather than developing in-house know-how and capability. Whether assessment was conducted in-house or outsourced, assessors often used rather arbitrary and subjective methods to categorize the level of risk rather than seeking out authoritative sources of information or consulting those with experience of the machinery in end use. Only a small proportion of firms involved their customers in risk assessment, and even fewer involved actual end users of machinery.

These findings about weaknesses in risk assessment by manufacturers and their consultants extend European studies, which have found that assessment is often inadequate (Boy and Limou, 2003, pp. 37–9, 62–3; Crabb, 2000, pp. 20, 39). Collectively, the findings raise serious questions about the prospects for machinery manufacturers institutionalizing arrangements to comply with their continuing legal obligations, and the emphasis to be placed upon risk assessment in legal instruments, and inspection and enforcement by safety regulators.

There is a challenge for regulators and policy makers, and industry and professional stakeholders, to consider how those involved in machinery design and construction can be encouraged and supported to develop the capacity to assess and manage risks effectively through the kind of experiential learning they favour. In particular, how can manufacturers learn what it means to assess and manage safety risks rigorously and systematically, from early in the life cycle of machinery? Can these practices be integrated with the trialling of prototypes and models, consultation with customers and end users, and other everyday activities of machinery design and construction? As with manufacturers' learning about machinery safety matters more generally, regulators could play a greater role in enforcing the need for manufacturers to develop capacity in assessing and managing risks, on an ongoing basis. They could also play a role in working with industry and professional stakeholders to develop experiential learning initiatives in which manufacturers have the opportunity to engage in constructive risk assessment and risk management activities. These issues are explored further in Chapter 9, which outlines how such learning initiatives could be usefully based on the practices shown in this research to be linked with better performance for substantive safety outcomes and feasible for firms of different sizes.

The book now turns to the final element influencing business responses – motivational factors. As foreshadowed in this and earlier chapters these factors, together with knowledge, shaped manufacturers' responses on machinery safety matters and their performance for hazard recognition, risk control and provision of safety information.

# Chapter 8
# Motivational Factors and Performance

The preceding chapters have established that manufacturers' actions on machinery safety matters were partially shaped by the knowledge base of those involved in design and construction activities. However, firms' performance could not be explained with reference only to knowledge. In their seminal work on compliance and enforcement, Kagan and Scholz (1984) recognized that as well as competence, motivations and attitudes played a role in shaping business responses and compliance. This chapter takes up the conceptual theme of motivational factors. It shows that differences between firms in their operations and interactions with external actors engendered diverse motivations which, together with the values and attitudes of the key decision makers in firms (the key individuals), shaped firms' responses on machinery safety matters.

The analysis here builds on safety and socio-legal scholarship, which has established the contextualized and plural nature of regulatees' motivations that shape their decision making and responses to regulation generally, and for safety in particular (Parker and Nielsen, 2011, pp. 9–14; Kagan, Gunningham and Thornton, 2011). Hale (2003) has argued that moral outrage at unnecessary suffering and premature death must be the basis for concern with health and safety matters. Yet, as discussed below, empirical studies of businesses' responses to safety and other social regulation demonstrate that legal, economic and social imperatives constitute the motivations that shape their actions, and a sense of moral agreement with regulatory goals is more rarely a factor.

For occupational health and safety (OHS) regulation, a study of UK firms concluded they were motivated by self-interest, which existed if poor safety standards had the potential to threaten business survival, if there were serious and well-recognized health risks, and/or when firms were large and highly visible to the inspectorate or local community (Genn, 1993). In the absence of such conditions, firms lacked self-interest and subordinated safety to more immediate production and profitability goals. In contrast, senior managers in Australian companies claimed to be motivated by legal, economic and reputational concerns, and a moral commitment to prevent harm (Hopkins, 1995, ch. 11). For Australian construction firms, profitability was the driving force behind their responses to work-related fatalities and, while influential and large firms were able to accommodate safety, those prone to competitive pressures chose between profit and safety (Haines, 1997, chs. 1, 7, 10). Managerial attitudes[1] also played a role, as the basis for interpreting and justifying action on safety as consistent with firm success, or not.

In the field of environmental regulation, research with paper pulp manufacturers in four countries found that legal, economic and social pressures, together with management style, including attitudes, determined firms' environmental performance (Gunningham, Kagan and Thornton, 2003, chs. 3–6). Firm size and sophistication also shape business motivations, as demonstrated in a study contrasting small and large US firms engaged in chemical processes (Gunningham, Thornton and Kagan, 2005). For small firms the principal motivations were state regulation and major customers, but for large firms risk management, reputation and the desire to maintain the trust of local communities were the key motivations. For small US trucking firms, general economic pressures coupled with a firm's market niche and financial condition influenced environmental performance (Thornton, Kagan and Gunningham, 2009).

Examining corporate management of social and legal responsibility more generally (Parker 2002, ch. 3) established the diversity of senior management's motivations. These included the potential for a competitive advantage, a sense of responsibility under a social contract, the need to garner good publicity and legitimacy in public eyes, and personal moral codes. Other scholarship suggests that divergent motivations may also stem from conflicting goals across regulatory regimes, where compliance in one regime impedes compliance in another (Haines and Gurney, 2003).

As the preceding empirical studies suggest, business motivations are of different types, which socio-legal scholars have characterized as legal, economic, social and normative, or a sub-set or amalgam of these (Ayres and Braithwaite, 1992, pp. 23–5; Gunningham, Kagan and Thornton, 2003, pp. 35–8; Kagan, Gunningham and Thornton, 2011; May, 2005; Parker and Nielsen, 2011, pp. 10–12). Legal motivations derive from the perceived authority of the law and threat of penalties if non-compliance is detected, while economic motivations relate to regulatees' commercial goals to maximize profit or material utility. Social motivations stem from regulatees' desire to earn the approval and respect of significant people with whom they interact (to be seen to do the right thing), and normative motivations arise from regulatees' desire to conform to internalized norms or beliefs about right and wrong. Reducing motivations to such broad categories is, however, a challenging activity precisely because of the highly contextualized and plural nature of motivations.

The significant contribution of this chapter is in providing a nuanced account of the complex nature of motivations, values and attitudes (motivational factors), and how their influence played out in manufacturing firms. These insights are of obvious interest to readers coming to this book from safety, socio-legal or sociological perspectives, as they further develop understandings of the contextualized and plural nature of the motivational factors that shape business responses to risk,

---

1   Haines distinguishes a virtuous and a blinkered culture based on managers' attitudes towards success, the role of various workplace parties and the OHS regulator (Haines, 1997, pp. 96–7).

and the relative importance of state regulation alongside other influences. The findings have wider relevance for specialists and practitioners in human factors and engineering, regulators and policy makers, any of whom may seek to secure positive business responses for safe design or other social goals, as they enhance understanding of motivational factors as drivers or barriers impacting on business performance.

In study firms, motivations derived from a variety of legal or quasi-legal imperatives, and firms' economic goals of ensuring the marketability of machinery and profitability. The values and attitudes of key individuals also mediated firms' responses on safety matters, providing assumptions about responsibility for machinery safety. Motivations, values and attitudes (collectively motivational factors) were capable of supporting sound machinery safety outcomes and providing firms with positive rationales for taking preventive action, or providing negative justifications for firms to limit or not take action. A recurring theme is the interplay of co-existing motivational factors, which may constitute mutually reinforcing rationales for a firm's action or inaction on safety matters, or push and pull a firm's decision making and action in different directions. Also revealed are the motivational factors linked with better or poorer performance for hazard recognition, risk control and safety information. The chapter conclusion raises some implications for those wanting to encourage attention to safety in machinery design and construction, highlighting the need to pay greater attention to the dynamics of motivational factors that affect commitment, or at least willingness, to perform well for substantive outcomes.

## Legal and Quasi-Legal Motivations

Although awareness of legal obligations for machinery safety was generally low in study firms (Chapter 4), and experience of inspection and enforcement was minimal and spasmodic (Chapter 5), various legal imperatives were among the motivations for some firms to take action on machinery safety matters. The perceived authority of some quasi-legal technical standards was also a source of motivation in some firms, as well as a non-specific concern about litigation or legal liability arising from unsafe machinery.

There were three firms in which the key individuals were sufficiently aware of the obligations in Australian OHS law for these to influence the firms' action on machinery safety. In a group of other firms the key individuals knew a little about these obligations and the law was the impetus for their firms to take some preventive action, but not to ensure that they were fully aware of and complied with their obligations. In total, there were 14 firms motivated by Australian OHS law (21 per cent (14/66)). The action taken by these firms included conducting a risk assessment for their machinery, or engaging a consultant to do this, although the methods applied in their assessments derived from Australian Standards (which generally did not have any formal legal standing)[2], as well as customers and other industry sources (see Chapter 7).

Most study firms referred to technical standards as part of the process of machinery design and construction (Chapter 6), but there were four firms for which a technical standard was a driving force for action on machinery safety (6 per cent (4/66)), and not only a source of information. The technical standards provided more detailed specifications and procedures for machinery safety and fitness for purpose, and the firms perceived the standards to have authority (Productivity Commission, 2006, pp. 6, 14; Wettig, 2002). Three of the standards driving manufacturers' action applied to particular types of machinery or equipment (boilers and pressure vessels, patient lifting equipment and cranes), and the fourth provided specifications and procedures for safeguarding machinery generally. The comments of the manager of a firm that manufactured patient hoists reflect the motivation to comply with a particular technical standard. He explained that:

> ... my predecessor who founded the company ... what drove him initially was not occupational health and safety legislation, it was the standard [ISO/EN 10535][3] and there was the need to market the product. And the standard had greater commercial value ... the commercial value of complying and making sure that your machine is safe for other people to use. (Manager, Manufacturer 39)

Each of the four firms motivated by a particular technical standard took action to comply with the standard and applied the standard closely, regardless of any legal standing it might have. Two of the standards were mandatory or evidentiary under Australian OHS law, and two did not have any legal standing. For these four firms standards had intrinsic value because they provided authoritative engineering principles and/or the firms could accrue commercial benefits by marketing their machinery as complying with the standard.

A further legal motivation for 20 study firms was inspection or enforcement by Australian OHS regulators (30 per cent (20/66)). These firms derived motivation from dialogue and negotiation with inspectors, receiving or being aware of notices affecting their machinery, being prosecuted or investigated for prosecution, having direct knowledge of a prosecution by reading a case transcript, and/or messages about inspection and enforcement amplified through their customers or distributors (see Chapter 5). The type of action taken by these firms included conducting a risk assessment or engaging a consultant to do so, making changes to

---

2    The standard for machinery, AS 4024 (Standards Australia, 1996) was part of an approved code of practice in Victoria, but had no legal standing in South Australia. The standard for risk management, AS/NZS 4360 (Standards Australia, 1999) did not have legal standing in either state (see also Chapter 2 for discussion of technical standards and Australian OHS law).

3    This was the international standard for *Hoists for the Transfer of Disabled Persons, Requirements and Test Methods*, ISO 10535 (ISO, 2002; updated in 2006).

risk control measures, redesigning machinery to eliminate hazards at the source, and/or providing or making changes to machinery safety information. Only one large firm had taken steps to improve its ongoing management of safety, including attention to safety in machinery design and construction. The firm took this action after it was prosecuted and because the firm's plans to gain status as a self-insurer for workers' compensation, which was financially advantageous to the firm, might have been harmed by the prosecution.[4]

For 11 firms (17 per cent (11/66)) that supplied, or intended to supply, their machinery in Europe, the motivation to take some preventive action was the regulatory regime implementing the *Machinery Directive* (European Commission, 1998).[5] If machinery did not comply with the essential health and safety requirements established in the *Machinery Directive*, these firms, or their agents in Europe, could have been prohibited from or restricted in placing the machinery on the market or into service in Europe (European Commission, 1998a, art 4, annex; see also Chapter 2). There was therefore a commercial incentive for these firms to address safety matters in order to avoid impediments to the supply of their machinery in Europe. The business goal of trading in Europe motivated a range of responses from firms including conducting a self-assessment for the firm's machinery or engaging a consultant to assess the machinery, applying relevant European harmonized standards and providing safety information.

The final legal motivation was what key individuals referred to as a concern about litigation or legal liability. This was a motivation for 19 firms to take some preventive action (29 per cent (19/66)).[6] They were worried about the possibility of some kind of legal action involving their firms and the adverse consequences for their businesses if such action was taken. The following comments capture the essence of this motivation:

> ... the way litigation is starting to creep into this country, we're very, very conscious that when we make a machine we look at it from all different angles, safety wise. (Technical services manager, Manufacturer 12)

> ... you're going to be responsible for providing that equipment and under the laws these days if you're negligent in any way it could end up in some form of litigation which could drag on and get very expensive. (Business development manager, Manufacturer 59)

---

4   A successful prosecution for a breach of the OHS legislation was grounds for declining a licence to self-insure under the workers' compensation scheme (WorkCover Corporation, 2001a, p. 16).

5   The *Machinery Directive of 1998* was in force at the time of data collection for this research. It was revised and reissued in 2006 for application in 2009.

6   The figures for firms concerned about litigation/legal liability do not include those with specific experience or knowledge of particular prosecutions under Australian OHS law. These firms are included in the discussion of enforcement as a motivation.

For study firms there was the possibility they could be involved in civil actions for the tort of negligence and, in at least some Australian jurisdictions, for breach of statutory duty, if a person sustained injury or damage arising from the firm's machinery (Davies and Malkin, 2003, chs. 1–6, 14; Luntz and Hambly, 2002, chs. 2–5, 10). There was also the possibility of a claim for damages for an alleged breach of contract, for example by a customer against the manufacturer (Clarke, Clarke and Courmadias, 2005; and see Australian Government, n.d.).

Two of the firms concerned about litigation or legal liability had been defendants in civil actions. One of these firms was sued by a worker who was injured when he accessed the interior space of the firm's machinery, which contained a rotating auger. The civil action motivated the firm to develop several new design features to eliminate the need for operators to enter the interior of the machinery to clear blockages or for maintenance, and also to make it difficult for any other person to access this space. The second firm, which had produced a particular type of machinery in large numbers for many years, had been involved in several civil actions. This prompted the firm to produce additional machinery safety information, conduct testing for the risk of projectiles emitted by the machinery and incorporate dead man's clutches to prevent the machinery from moving if the operator was not actively controlling it. Whether a firm had direct experience of civil actions, or only a generalized concern about the possibility of such actions, the potential for litigation or legal liability was a motivation for some firms to seek out and address sources of harm, and to make machinery safer. It was also a motivation for firms to apply technical standards, or to provide or improve machinery safety information.

Overall one or more legal or quasi-legal imperatives constituted motivations for 44 study firms to take some form of preventive action (67 per cent (44/66)). These motivations stemmed from the desire to avoid an adverse outcome, such as the cost of legal action, or to secure a positive outcome, such as being able to market machinery as complying with particular legal requirements or technical standards. Legal and quasi-legal motivations were reasons behind some firms' decisions and actions on machinery safety matters, whether they were grounded in personal experience or just the perception of a general threat or benefit, and whether or not they were underpinned by specific knowledge of legal obligations or types of legal action. These motivations variously prompted firms to conduct a risk assessment or engage a consultant to do so, apply technical standards, redesign machinery or enhance risk control measures, or provide or make changes to safety information, and through these actions they had the potential to influence performance for substantive safety outcomes.

Specific legal or quasi-legal motivations did not operate alone. They co-existed with other such motivations, and/or with economic motivations, or values and attitudes. In a particular firm there was interplay between different motivational factors which might mutually reinforce the need for action on machinery safety matters or might be inconsistent, pulling firms' efforts in different directions, as discussed further below.

**Economic Motivations**

A series of economic motivations influenced study firms' action on machinery safety matters. These economic motivations were derived from the business goal of ensuring the marketability of their machinery and firm profitability. While economic motivations were strong influences in some firms, only some types of economic motivations encouraged constructive preventive action. Others had the potential to impede such action.

*Economic Motivations with the Potential to Support Action on Machinery Safety*

Four large manufacturers proactively managed business risks and had more developed systems for self-regulation and corporate governance (6 per cent (4/66)). These firms integrated the management of safety, quality and other business risks, and adopted different approaches to safety management. One firm applied Australian Standards for safety management (Standards Australia, 1997; 2001b), and the other three implemented safety management standards as self-insurers under a state workers' compensation scheme (WorkCover Corporation, 2001a, pp. 9, 15–16). For the latter three firms there was an added financial incentive to proactively manage safety in order to protect their self-insurer status. An example was a firm that designed and constructed major plant installations. The firm's safety manager explained the relationship between safety, governance and economic motivations:

> ... due diligence and corporate governance are factors that have driven health and safety ... In terms of our design, it's much easier to at least have a conversation about health and safety now ... I think a lot of that has been on this due diligence, corporate responsibility that has driven that. ... We maintain our self-insured license by compliance with the [WorkCover] standard. Maintaining exempt status saves us around about a half to three quarters of a million dollars a year. When our profit margin is only around $1.5m that's 50 per cent of our profit that we would lose if we lost our self-insured status. So that's a fairly powerful motivating factor ... . (Safety manager, Manufacturer 15)

The economic motivation to minimize business risks drove the four firms' systematic management of safety. In turn, their systematic approach supported more rigorous hazard recognition and decision making about risk control. Like the large chemical companies in Gunningham, Thornton and Kagan's (2005) study above, the motivation of the four large firms to take preventive action primarily related to risk management, with management systems playing a role.

Three other firms produced safe machinery as a business opportunity (5 per cent (3/66)). They marketed their machinery as safety or ergonomic solutions and addressed these features of their machinery as an integral part of their business operations. One firm had developed a hand-held, self-supporting drill to minimize upper limb injuries. The firm customized the drill to order, providing solutions to ergonomic problems in drilling and for handling loads. A second firm produced a new type of surface finishing machine to resolve the hazardous dust, ergonomic and workload problems of conventional machines. The managing director explained:

> It definitely does sell the machine over other machinery with these safety features. So it's been worth our persistence in just being safe and putting those safety things in. If we could find any more we'd put them in as well because … the end user is interested in the safety features. (Director, Manufacturer 10)

For the three firms producing safe machinery as a business opportunity, safety features were marketable and they were motivated to seek out safety problems and find solutions to them. These firms took preventive action because they perceived that safety was integral to the future success of their businesses (see also Haines, 1997, chs. 6, 7, 10).

Other firms were motivated by concerns about machinery quality. In addition to the four firms with systems for managing safety together with other business risks, there were another six firms in which quality was a motivation to take action on machinery safety matters. In total, ten firms linked machinery safety and quality (15 per cent (10/66)). These were not the only firms in which quality was a goal but they were the only ones in which quality was a motivation to act on machinery safety, and for which safety was integral to quality. The technical services manager of a firm that manufactured compressors explained the link between machinery safety and quality, and how the firm recognized and addressed safety problems through its quality improvement arrangements:

> … it's a tremendously powerful system. It's pushed this company to where it is today by listening to what goes on out in the field. … The QIR [quality improvement report] tells us right now that we've got a problem … We had a structural problem in the design of these hoses … we stopped the production line … The first thing we did was then go through all the design criteria and so forth. There's an element of danger here, if it burst off and blew hot oil and so forth around there, that's one of the drivers, but really to be putting product off our line that wasn't suitable for field use was – you just can't do that and really that's for our own responsibility and certainly our market because the perception of our market and so forth out there and under ISO [the quality standard] we could never do that … So this line shut down for a fortnight and the reason that happened was that you just can't do these things overnight. You've got to redesign, then you've got to test … that was driven out of our quality improvement system. (Technical services manager, Manufacturer 60)

As this example illustrates, in firms that linked machinery safety with quality, concern about quality was a motivation to take action on safety matters. These firms obtained and applied relevant technical standards, had arrangements for seeking feedback about problems in end use, including hazards and hazardous exposures, and took action to address these problems.

Another motivation for ten firms to take action on machinery safety was competition. This was manifest as the need to be competitive with other suppliers of the same type of machinery, or the desire to secure a competitive advantage over other suppliers (15 per cent (10/66)). As the engineering manager with a firm that manufactured packaging machinery stated:

> ... we do tend to up our category rating on our safety as a selling point. We're now making case packers which contain robots in them and a lot of people are very cautious of the robots so we've upped our category rating. ... it does drive us a little bit because it is another selling point that we can use to be able to compete with our competitors. (Engineering manager, Manufacturer 13)

This example is typical of the firms in which competition was a motivation for preventive action. For these firms safety features were marketable, as they perceived that their machinery was more competitive in the market place if they enhanced the safety features.

For eight manufacturers, their customers or distributors were the motivation for action on machinery safety (12 per cent (8/66)). While interaction with customers and distributors was an integral part of doing business as a manufacturer, in these eight firms customers or distributors drove their action on safety matters. Six of these were small firms. The size of the other two varied from small to medium with labour hire workers used in busy production periods. All eight firms supplied their machinery to large customers or distributors which imposed safety requirements, sometimes in the form of detailed machinery specifications. The firms complied with these requirements because they produced the machinery under contract and it was in their commercial interests to do so. This perspective is illustrated in the following comments:

> ... for us to be successful to win contracts we had to meet those requirements ... we hate doing some of the safety requirements they ask for – we think they're over the top and we don't necessarily agree with some of them – they add cost to the machine ... but over the years we've naturally learnt what they want. (Managing director, Manufacturer 5)

> ... the people that we're supplying it to ... they walk through, they look at it and check it out. They may make some recommendations ... There's a strict set of procedures that are followed to ensure that the product is safe. The final safety goes to the customer. The customer approves or disapproves ... (Manager, Manufacturer 17)

The supply chain pressure reflected in these examples has been more widely recognized as a motivation, especially for smaller firms with large customers

(Gunningham, Thornton and Kagan, 2005; Hutter 2011, pp. 315–17). Indeed, some safety and socio-legal scholars have suggested that limitations in the state's ability to regulate certain economic relationships directly may be compensated for by encouraging large enterprises to act as third party, surrogate regulators, for example in sub-contracting relationships and with small business (Gunningham and Johnstone, 1999, p. 133; Gunningham and Sinclair, 2002, pp. 17–18; Hopkins and Hogan, 1998; Lamm and Walters, 2004; Walters, 2001, pp. 52, 375–6; 2002, pp. 45–6). The present research suggests that the influence of large customers or distributors did not lead to manufacturers having a greater commitment to machinery safety or taking sustained action on these matters. The firms whose action on safety was driven by customers or distributors did not take responsibility for machinery safety themselves. They perceived that their customers or distributors were taking responsibility for safety and, while they met the requirements imposed, they did so with a degree of ambivalence. They might regard a particular requirement as unnecessary or imposing an additional cost, and as action that they would not take unless required to do so for a particular contract.

A further motivation in seven firms was concern about reputation; either the manufacturer's or the reputation of the machinery (11 per cent (7/66)). These firms wanted to avoid bad publicity which they perceived would impact adversely on sales and put the survival of the business at risk. For some firms this was a motivation to recognize and control hazards before machinery was supplied. For others it was a motivation to address problems promptly, when they were reported after supply. As the managing director of a firm that manufactured agricultural machinery stated:

> Your reputation in the market place is very important. Public relations and so on is a very important part of being successful. So if it gets around the bush that we make something that's unsafe, we're not going to be there for very long ...
> (Managing director, Manufacturer 20)

The findings about the motivation engendered by reputational concerns resonate with other safety and socio-legal studies which have identified reputation as a motivation for firms to take preventive action (Genn, 1993; Gunningham, Thornton and Kagan, 2005; Hopkins, 1995, pp. 163–4; Hutter, 2011, pp. 140–6). In some contexts reputational concerns could be a form of social motivation, reflecting a desire to earn the approval and respect of significant others, and be seen to be doing the right thing (Parker and Nielsen, 2011, p. 11). Among study firms there was no evidence that reputational concerns and a fear of negative publicity sprang from anything other than commercial goals and the desire to ensure business survival.

In summary, a series of economic motivations provided firms with rationales for taking constructive action on machinery safety matters, although the scope and nature of the action prompted by economic motivations varied. Also, as demonstrated in the next section, not all economic motivations supported action on machinery safety matters. Some economic motivations provided firms with justifications for limiting or not taking preventive action.

*Economic Motivations with the Potential to Impede Action on Machinery Safety*

An economic motivation that adversely affected risk control action by half of the study firms was concern about machinery functionality (50 per cent (33/66). This was the concern to ensure that machinery would function well for its intended use. Firms with such concerns elected not to incorporate particular risk control measures, or not to increase the level of risk control, if they perceived that these measures would reduce the functionality of their machinery. The following comments, from manufacturers of agricultural machinery and of food processing systems respectively, capture the essence of the concern about functionality:

> ... a machine we invented which is a pastoral topping machine for open paddocks – the whole back of the machine is flared up at a 30 degree angle ... it throws it [cut grass] up to 30 feet in the air ... highly dangerous if you're standing behind it ... but as far as the purpose of that machine it must be open on the back. Also we manufacture machines that put mulch underneath vine rows and up underneath trees. Now if it's in a vine row the side of the slasher is open on the mower and it allows you to throw out to the vine line. Again you wouldn't want to be standing there ... .(Owner, Manufacturer 19)

> When you reach a point in the design where you say, "well if we guard this we'll render it inoperative, if we don't guard it, it's a safety problem", then we get our customer involved and say, "here's the position we're in, what can we do, is it possible for you in some way to have people holding two buttons while this machine is operating so that their hands are out of the way, or is it possible for you to train your people?" ... Not transferring the risk to them but examining the project and saying, "here are the problems" and "how are we going to overcome them", and there's generally a way of getting through it but you may not end up with an inherently safe piece of equipment. (Managing director, Manufacturer 57)

As these examples illustrate, functionality imperatives influenced some manufacturers' choice of risk control measures and limited the action that they took to control risks. This included firms not incorporating certain control measures, providing safeguards as optional extras to be purchased at the customer's discretion, or relying on less safe measures such as warnings to customers to keep clear of the machinery or to follow safe operating procedures. Alternatively, firms might negotiate with their customers to take certain protective measures after the machinery was supplied. In these firms, any impediment to the functioning of the machine constituted an adverse impact on its competitiveness with other similar machinery on the market, or its acceptance by customers. These findings reinforce

UK research with manufacturers of agricultural machinery, which found that a fundamental driver for these firms was to design machinery to carry out the task and meet the demands of the marketplace (Crabb, 2000, p. 1).

Another economic motivation specifically limiting risk control action was concern about cost. While cost was a consideration for study firms generally, some firms made choices about risk control measures in which cost was an over-riding consideration, even if machinery safety was compromised as a result (36 per cent (24/66))). In their view machinery must be produced within the resources available to the firm, cost must be contained to avoid loss of sales to cheaper competitors or, for custom-made machinery, the cost must be within the price set by the customer. Cost containment meant that some firms did not incorporate safe place controls that they deemed too costly, or they relied on safe person measures such as warning signs or advising customers to use safe work practices. Some firms provided certain risk control measures as optional extras or produced alternative models, one less safe and one safer, and their customers could choose to purchase the safer model at an additional cost. For custom-made machinery, some firms negotiated with their customers to implement control measures after the machinery was supplied, rather than the manufacturer designing or building in safety features.

An example of a firm motivated by containing costs was a manufacturer of agricultural machinery, which had developed an innovative design solution to address a significant hazard after several fatalities and serious injuries involving the firm's machinery, and inspection and enforcement by OHS regulators in several states. Even with these pressures to take effective preventive action, the firm was reluctant to produce a model incorporating the design solution unless this could be done without the new model being more costly than the machinery supplied by the firm's competitors. The managing director explained:

> … we see it as a significant competitive advantage in this feature, especially if we can do it at cost neutral, compared to our competitors. If the cost is significantly higher then it doesn't become a safety feature. Unfortunately in the market that we're working in safety doesn't sell. If, however, safety is there at the same price then it will sell. (Managing director, Manufacturer 62)

This example illustrates how cost was an over-riding concern for some firms, which they felt justified them not implementing safety measures even when they were warranted by the magnitude of the risk. The cost containment perspective was clearly out of step with manufacturers' legal obligations. The reasonably practicable standard in Australian OHS law required manufacturers to weigh the magnitude of the risk against the cost of implementing particular preventive measures, and to take those measures unless the burden was grossly disproportionate to the risk (Bluff and Johnstone, 2005). In Europe, manufacturers were only entitled to put machinery on the market if it was not a danger to health or safety, and they could choose whether to comply with essential health and safety requirements, or to comply with relevant harmonized standards in lieu of these (see Chapter 2).

A further impediment to some study firms taking action on machinery safety matters was their view that there was an uneven playing field in the market for their machinery because of different expectations by OHS regulators, within or between Australian states, or for locally produced compared with imported machinery. Concern about inconsistent inspection and enforcement was an impediment to nine firms taking risk control action (14 per cent (9/66)). The underlying motivation for these firms was economic rather than legal as they perceived that if they incorporated particular risk controls when their competitors were not expected to do so, their own machinery would be less competitive in the market.

An example was a press manufacturer which, in the context of inconsistent inspection and enforcement by OHS regulators in several Australian states, chose to continue to sell an unsafe model of machinery, despite having developed a safe design solution (see Chapter 5). The firm believed that it would lose sales to competitors if it marketed only the safe design model because, as it perceived, the firm's competitors had not been required by OHS regulators to meet an equivalent standard. A second firm highlighted variation in inspection and enforcement of their machinery by different OHS regulators. The regulator in one Australian state had required the firm to produce a full guard so that an operator could not access the danger zone of the machinery, in any circumstances, while the regulator in another state had accepted a narrower but higher guard, and regulators in two other states had requested further variations on guarding the same machinery.

The impact of inconsistent inspection and enforcement on compliance with Australian OHS law was highlighted in the case of *Lyco Industries*[7] in which Schmidt J called on the OHS regulator in the state of New South Wales to deal with continuing and serious threats to safety brought about by other firms continuing to sell the same kind of machinery in an unguarded condition. The problem of firms being less inclined to comply if they perceive that others are getting away with non-compliance has also been recognized in other areas of social regulation (Gunningham, Thornton and Kagan, 2005). For machinery manufacturers, the uneven playing field arising from inconsistent inspection and enforcement provided a rationale for failing to adopt safer designs, incorporate safer risk control measures or take unsafe models off the market.

*Summary of Economic Motivations*

One or more economic goals constituted motivations for 58 study firms (88 per cent (58/66)). These motivations were intrinsically linked with firms' concerns about the marketability of their machinery and firm profitability. The cost of compliance is well recognized by socio-legal scholars as a factor likely to limit compliance with regulation due to organizations' pre-occupation with short term

---

7   *Inspector Ruth Buggy v Lyco Industries Pty Ltd* [2005] NSWIRComm 423, [39].

profitability (Gunningham and Johnstone, 1999, p. 69; Gunningham, Thornton and Kagan, 2005). For study firms, cost was only one of a series of economic motivations shaping their responses on machinery safety matters. Only some of these economic motivations provided firms with positive rationales for taking preventive action. Others provided firms with justifications for limiting or not taking action.

In the operations of a particular firm, motivations reflecting different economic goals might co-exist, reinforcing positive rationales or negative justifications, or requiring firms to reconcile conflicting goals. For example, in one firm there was a synergy between goals relating to machinery quality and securing a competitive advantage that provided the firm with strong rationales for taking action on machinery safety. In another firm, over-riding concerns about machinery functionality and cost provided strong justifications for limiting or not taking preventive action. In a third firm, recognition that certain safety features could help secure a competitive advantage were balanced precariously against concerns that these measures might affect the functionality of the machinery.

In most firms, particular economic motivations combined with other economic motivations, with legal or quasi-legal motivations, or with the values and attitudes of key individuals (80 per cent (53/66)). The mix of economic and other motivational factors in a firm influenced that firm's overall response on machinery safety matters. Like economic motivations, values and attitudes (as discussed below) could either encourage or discourage action on machinery safety, and reinforce or counter motivations supporting preventive action.

## Values and Attitudes as Motivational Factors

### The Moral Obligation to Protect Human Health and Safety

Values constitute motivational goals and guide the way individuals evaluate situations, select actions, and explain or justify their evaluations or actions (Licht 2008, pp. 3, 41). They are deep-seated, serve as individual standards or criteria for behaviour, and reflect matters that are desirable means or ends in themselves (Glendon, Clarke and McKenna, 2006, p. 189; Licht, 2008, pp. 21–2; Reber and Reber, 2001, p. 783). They are akin to normative motivations which arise from regulatees' desire to conform to internalized norms or beliefs about right and wrong (Kagan, Gunningham and Thornton, 2011). One value type is security which encompasses safety (Licht, 2008, pp. 56–7; Schwartz and Bilsky, 1987), and was reflected in the moral obligation to protect human health and safety expressed by key individuals in some firms. In principle, this value might be expected to support action on machinery safety matters.

In nine firms, the key individuals stated that such a moral obligation to protect human health and safety motivated them to take action on machinery safety (14 per cent (9/66)). They expressed this as a matter of conscience; an ethical responsibility to ensure that the firm's machinery did not hurt people. It was the view that it would be morally reprehensible for the manufacturer to knowingly supply machinery that was unsafe, and that the key individuals concerned would not want to live with the knowledge that someone was hurt because they had not addressed machinery safety matters. As the managing director of a firm that manufactured metal processing machinery stated:

> I think the biggest one clearly is, for me anyway, the potential of someone getting hurt, or worse, getting killed. Fundamentally it's a conscience issue there for us, before any other considerations are taken into account. (Managing director, Manufacturer 40)

Those expressing a moral obligation to protect health and safety suggested it was a desirable end in itself. Five of the nine firms in which key individuals expressed this moral obligation also had legal or economic motivations that supported action on machinery safety. In these firms, the normative motivation reinforced and provided further justification for the firms' action on safety matters. The other four firms were preoccupied with at least some commercial goals that did not support action on machinery safety. In these firms, the espoused normative motivation had to compete with the firms' other priorities. There was a trade-off between the moral obligation to protect people from harm and the firms' economic motivations.

## The Unsafe Worker Attitude

While a sense of moral obligation provided a justification for manufacturers to take action on machinery safety, some attitudes of key individuals provided justifications for limiting or not taking preventive action. Attitudes are tendencies to act in a consistent way toward a particular object or situation, and they influence behaviour via motivation (Glendon, Clarke and McKenna, 2006, pp. 187–9; Reber and Reber, 2001, p. 783; Sundström-Frisk, 1996; 1999). They enable reinforcement of individual desires and requirements, provide a defence mechanism to protect individuals from harsh realities, and are a means for individuals to order, make sense of and react consistently to the world around them (Glendon, Clarke and McKenna, 2006, p. 207). While attitudes are less deep-seated than values, they are not readily altered and a person is only likely to question and change a particular attitude when it is no longer useful. Among key individuals in study firms there were two key attitudes, both with negative consequences for machinery safety.

The first was the unsafe worker attitude; the perspective that machinery users act unsafely with machinery. Key individuals expressed this attitude in a little over half the study firms (55 per cent (36/66)). According to this attitude users were at fault or to blame for injury or incidents involving machinery, acted foolishly when interacting with machinery, or actively disarmed or removed safeguards. The following examples, from manufacturers of slashers and food processing machinery respectively, capture the essence of the unsafe worker attitude:

> ... you can think of anything you like and you're always going to miss something, like someone can go and try to lift it [the slasher] up at 90 degrees and start pruning their hedge. That is really not what they're designed for but if they really wanted to I suppose they could ... No matter what you do, you're sort of damned if you do and damned if you don't to a certain degree ... it doesn't stop someone from laying down on the ground and putting their arm under it either. (Managing director, Manufacturer 3)

> The operator has got to have some responsibility for his own safety. Whereas if we had a machine that was totally safe ... he then becomes totally reliant on the machine never being able to do something to him ... when we do put safety devices onto things, normally the operator will find a way of bypassing it, because it slows him down, or it stops the machine. (Director, Manufacturer 36)

For key individuals with a negative unsafe worker attitude, the possibility of unintended use was not a motivation to change machinery design to minimize this potential and the consequences of it. There was little recognition by these key individuals that users might simply make mistakes through faults in the machinery, error, fatigue, to get the job done with less strain, or for other reasons, or that changes to design could reduce the potential for unintended use (for literature on this see Chapter 6).

In general, the key individuals expressing the unsafe worker attitude did not look beyond the potential for unintended use to identify underlying reasons for it, or take steps to have their firms develop safe design solutions. Rather, the unsafe worker attitude provided them with the rationale that whatever the firm did to address machinery safety matters, it would not be enough. Firms in which this attitude prevailed continued to use safeguards they knew could easily be removed or disarmed, or relied on cautions, warnings or other less reliable safe person measures.

The unsafe worker attitude is predicted by attribution theory (Dejoy, 1994; Glendon, Clarke and McKenna, 2006, pp. 83–91), according to which individuals make inferences about the causes of events. These causal attributions more strongly influence actions than actual causes, and there is a strong tendency to over-emphasize individual responsibility when judging the behaviour of others. This fundamental attribution error operates in work situations as much as in everyday life, and is manifest in the emphasis often placed upon the role of workers, rather than situational factors, in workplace incidents (Dejoy, 1994; Glendon, Clarke and McKenna, 2006, pp. 88–9; Kouabenen, et al., 2001).

In her study of work-related fatalities in Australian construction firms, Haines found that key individuals attributed particular deaths to the workers who died, thereby rationalizing these events as beyond their control (Haines, 1997, ch. 5). She concluded that blaming the worker was a coping strategy, which enabled key individuals to externalize responsibility for workplace deaths but did not explain differences in firms' responses to these events.

The unsafe worker attitude played a more defining role in shaping study firms' responses on safety matters. The attitude provided key individuals in some firms with a defence mechanism against the harsh reality that unsafe machinery produced by their firms had caused fatal or serious injuries. In addition, whether or not key individuals were aware of hazardous incidents involving their firms' machinery, the unsafe worker attitude provided them with a justification for the fact that their firms continued to produce and supply machinery that they knew to be hazardous. The attitude underpinned assumptions that framed manufacturers' decision making and action. In the context of economic motivations concerning machinery functionality, cost and competitiveness, the unsafe worker attitude provided key individuals with a justification for not taking steps to redesign their machinery and for not incorporating particular risk control measures.

Attribution of fault to end users enabled key decision makers to avoid responsibility and precluded them from the need, at least in their own minds, to take more rigorous preventive action. Such attribution is an impediment to safety (Hasle, Kines and Andersen, 2009), but it is likely to prove to be an intractable influence on manufacturers' decision making and action, unless its usefulness in underpinning firms' economic motivations can be successfully challenged.

## The Attitude that Safety Has Gone Too Far

A second negative perspective expressed by key individuals in some firms was the attitude that safety has gone too far (12 per cent (8/66)). This was the perspective that risk is inevitable and that there is too much pre-occupation with safety in society generally. The comments of the owner of a firm that manufactured cranes, hoists and other machinery for use in motor vehicle repairs were typical of this attitude. He stated:

> It's got to a point where it's become ridiculously ridiculous [sic] on what you have to go through to try and eliminate risk … Where do you draw the line? It's got to the point in some of the cases I've seen in some of the reviews where you think it's time to basically pack up, shut up and go home because you cannot cover everybody all of the time, but if you do get pinched or someone takes you – well it's good night nurse, why bother? (Owner, Manufacturer 4)

Key individuals with the attitude that safety has gone too far perceived that risk is a part of life, and that it is not possible to protect all people from all risk, all of the time. According to this perspective it was unreasonable to expect manufacturers to eliminate or minimize risk when, in the individual's opinion, it was not warranted. Rather, those with this attitude suggested that safety should be a matter of common sense. In firms in which key individuals expressed the attitude that safety has gone too far, it co-existed with the unsafe worker attitude and provided key individuals with an additional rationale for limiting or not taking action on machinery safety matters.

*Summary of Values and Attitudes as Motivational Factors*

Values and attitudes acted as a lens through which manufacturers' decision making and action on machinery safety was interpreted and shaped. They mediated firms' responses on safety matters by providing assumptions about responsibility for machinery safety. A positive, moral obligation to protect human health and safety was less common (key individuals in 14 per cent of firms), than the negative unsafe worker attitude (key individuals in 55 per cent of firms). The unsafe worker attitude also sometimes co-existed with the attitude that safety has gone too far (key individuals in 12 per cent of firms).

In 41 firms, different combinations of values or attitudes co-existed with legal, quasi-legal or economic motivations (62 per cent (41/66)). There was considerable potential for the more prevalent negative attitudes to impede responses on machinery safety by reinforcing firms' justifications for limiting or not taking preventive action on grounds of functionality, cost, avoiding a competitive disadvantage, or for other reasons. On the other hand, the less commonly expressed moral obligation to protect health and safety might reinforce the need for preventive action in firms with positive economic or legal rationales for addressing safety matters, but could be over-ruled by economic motivations that favoured machinery marketability and firm profitability over safety.

The combination of, and interplay between, motivational factors in the operations of a particular firm and its interactions with external actors, shaped how that firm responded on machinery safety matters, and the firm's overall performance for the substantive safety outcomes. Certain factors tended to support better performance for the substantive outcomes. The link between motivational factors and performance is the focus of the next section.

## Motivational Factors Linked with Better or Poorer Performance

So far this chapter has shown that some motivational factors encouraged or prompted manufacturers to take constructive action on machinery safety such as conducting a risk assessment, applying technical standards, redesigning machinery, improving risk control measures, receiving and acting on feedback about problems with the machinery, and providing or improving machinery safety information. Other motivational factors deterred firms from taking these types of action.

While the constructive motivational factors had the potential to sustain good performance for substantive safety outcomes, the deterrent factors potentially contributed to poorer performance. This section demonstrates that firms that accepted responsibility for machinery safety, and had positive rationales for addressing safety matters, tended to perform better for the substantive outcomes. On the other hand, firms with motivational factors consistent with not accepting and taking responsibility for machinery safety tended to perform more poorly for these outcomes.

The approach to analysis involved systematically reviewing the data about the performance of firms in which particular motivational factors held sway,[8] as the basis for reflecting on plausible relationships between levels of performance and particular factors, and inductively developing explanation which accounted for differences in performance for the substantive safety outcomes. Of particular interest were motivational factors linked with markedly better or poorer performance, when compared with the performance of firms in the sample overall. As demonstrated in Chapter 3, of the 66 study firms 30 per cent had comprehensive hazard recognition; 14 per cent had a blinkered focus on mechanical hazards; 47 per cent used safe place controls as the primary risk control measures; 17 per cent used some advanced or innovative safe place controls; and 24 per cent provided substantial, good quality safety information. In order to distinguish more distinct trends, the focus here is on motivational factors for which the proportion of firms performing at a particular level, for a specific substantive safety outcome, was *at least 10 per cent above or below* the proportion of firms performing at that level in the sample overall.

## *Legal and Quasi-Legal Motivations, and Substantive Outcomes*

The performance of manufacturers influenced by legal and quasi-legal motivations differed with specific motivations variously linked with better performance for some substantive safety outcomes, or with similar or poorer performance compared with the sample of firms overall. The motivations linked with better performance for some outcomes were the Australian and European legal obligations for machinery safety, and quasi-legal technical standards.

Key individuals' awareness of machinery safety obligations in Australian OHS law was limited (Chapter 4), and only 14 firms were motivated by this body of law to take some action on machinery safety matters. Nonetheless, the firms *motivated by Australian OHS law* performed better than firms in the sample overall across several substantive safety outcomes. As set out in Table 8.1 (p. 154), the firms motivated by obligations in Australian OHS law were more likely to have complete hazard recognition (43 per cent (6/14)), and none were blinkered in their hazard recognition (0 per cent (0/14). These firms were also more likely to use safe place controls as the primary risk control measures (71 per cent (10/14)), and to provide substantial, good quality safety information (43 per cent (6/14)). They were, however, less likely to use more advanced or innovative safe place controls (7 per cent (1/14)). This finding does not imply that Australian OHS law was the cause of the higher performance for particular substantive outcomes. Rather, firms

---

8   The analysis focused on motivational factors for which there were ten or more study firms in which the particular factor applied. For these factors, the proportion of those firms performing at a particular level was calculated. For factors with influence in fewer than ten firms, broad trends in the data are discussed.

*Safe Design and Construction of Machinery*

**Table 8.1** **Legal motivations linked with better or poorer performance for substantive safety outcomes**

| | Substantive safety outcomes | | | | | | | | | |
|---|---|---|---|---|---|---|---|---|---|---|
| | Comprehensive | | Blinkered | | Safe place emphasis | | Advanced/ innovative | | Substantial, good info | |
| | n/N | % | n/N | % | n/N | % | n/N | % | n/N | % |
| Aus OHS law | 6/14 | 43 | 0/14 | 0 | 10/14 | 71 | [1/14] | [7] | 6/14 | 43 |
| EU law | 6/11 | 55 | | | | | 3/11 | 27 | 4/11 | 36 |
| Whole sample | 20/66 | 30 | 9/66 | 14 | 31/66 | 47 | 11/66 | 17 | 16/66 | 24 |
| Litigation | 4/19 | 21 | | | 7/19 | 37 | | | | |

*Note:* In Table 8.1, N is the number of firms in which a particular motivation applied, and n is the number of those firms performing at the stated level for the particular substantive safety outcome.

that accepted OHS law as a legal imperative had a positive rationale for taking action on machinery safety matters.

The firms *motivated by the European regulatory regime* for machinery safety had better performance for some substantive safety outcomes but did not perform markedly differently on other outcomes, compared with the sample overall. As set out in Table 8.1 above, these firms were more likely to have comprehensive hazard recognition (55 per cent (6/11)), and they were a little more likely to use advanced or innovative controls (27 per cent (3/11)). They were also more likely to provide substantial, good quality safety information (36 per cent (4/11)). These findings are consistent with the possibility that firms motivated by European requirements paid more attention to hazard recognition and machinery safety information, and/ or that the conformity assessment conducted by these firms, or their consultants, supported better performance for these outcomes.

The motivation linked with poorer performance for some safety outcomes was the non-specific concern about litigation or legal liability. As set out in Table 8.1 above compared with the sample overall these firms were less likely to have comprehensive hazard recognition (21 per cent (4/19)), and less likely to use safe place controls as the primary risk control measures (37 per cent (7/19)). Their performance for the use of advanced or innovative risk controls, and provision of safety information was similar to the sample overall.

The firms concerned about legal liability or litigation were not aware of the specific types of legal action, and did not know what action they should take to minimize the risk of legal actions against them. As regulatory scholars have concluded elsewhere, the threat of legal liability may not sufficiently motivate firms to improve safety outcomes if they are unaware of the findings of legal actions (Baram 2006, pp. 14–17; Haines 1997, p. 183). Firms might also respond to the threat of litigation with other loss mitigation strategies such as insurance, or indemnification clauses in contracts seeking to transfer responsibility to other parties rather than by overhauling and ensuring the efficacy of machinery safety measures.

The four manufacturers for which particular technical standards were the motivation for the firm's action on machinery safety matters tended to perform better for substantive outcomes. An exception was the use of more advanced or innovative controls, as only one of these firms used such measures. (As technical standards were only a motivation in four firms, proportions are not included in Table 8.1).

The performance of the 20 manufacturers influenced by inspection or enforcement by an Australian OHS regulator was similar to the sample overall for comprehensive hazard recognition (30 per cent (6/20)), a blinkered response (15 per cent (3/20)), use of safe place controls as the primary risk control measures (45 per cent (9/20)), use of advanced or innovative controls (25 per cent (5/20)), and provision of substantial, good quality safety information (25 per cent (5/20)). As discussed in Chapter 5, OHS regulators' involvement with machinery designers or manufacturers was infrequent and typically event-based, and focused on securing one-off compliance. The nature and infrequency of inspection and enforcement was insufficient to prompt firms to take more sustained and systematic action to ensure better performance for hazard recognition, risk control and provision of safety information. (As firm performance was not markedly different from the sample overall the trends for inspection and enforcement are not included in Table 8.1).

It is noteworthy, however, that among the five firms that used more advanced or innovative control measures, and were influenced by inspection or enforcement, it was their experience of prosecution under Australian OHS law, investigation for prosecution and/or inspectors' enforcement notices which prompted them to develop safe design solutions to significant risks with their machinery. The use of advanced or innovative control measures by these firms was their way of reconciling the regulatory pressure for greater risk control with their machinery functionality goals, which could only be achieved by redesigning their machinery.

*Economic Motivations and Substantive Outcomes*

Some economic motivations provided firms with positive rationales for taking action on machinery safety matters and these firms tended to perform better on at least some substantive outcomes. These were the motivations relating to quality, competitiveness, managing business risks and business opportunities.

The ten firms *motivated by quality concerns* performed better than the sample overall for all substantive safety outcomes. As set out in Table 8.2 below, they were more likely to have comprehensive hazard recognition (70 per cent (7/10)) and none were blinkered in approach (0 per cent (0/10)). They were also more likely to use safe place risk controls as the primary measures (80 per cent (8/10)), incorporate some more advanced or innovative controls (30 per cent (3/10)), and provide substantial, good quality safety information (50 per cent (5/10)).

The ten firms for which action on machinery safety was *motivated by competition*, performed better than the sample overall for some substantive outcomes. As set out in Table 8.2 below, none of these firms were blinkered (0 per cent (0/10)), and they were more likely to use safe place controls as the primary risk controls (60 per cent (6/10)), and advanced or innovative risk controls (30 per cent (3/10)).

The other positive economic motivations, managing business risks and business opportunities, were drivers in only a small number of study firms and so it is only possible to report trends for these. The four firms motivated to manage a series of business risks tended to have better performance for all substantive outcomes except advanced/innovative controls. The three firms that produced safe machinery as a business opportunity tended to have better performance for all substantive outcomes except safety information.

Considered together, the 21 firms that linked action on machinery safety with managing business risks, quality, competitiveness and/or business opportunities defined machinery safety as compatible with their commercial goals. They regarded the safety of their products as *integral to the success of their businesses*, or at least a key priority. These firms had sound commercial rationales for addressing machinery safety matters, which supported better performance for substantive

**Table 8.2    Economic motivations linked with better or poorer performance for substantive safety outcomes**

| | Substantive safety outcomes | | | | | | | | | |
|---|---|---|---|---|---|---|---|---|---|---|
| | Comprehensive | | Blinkered | | Safe place emphasis | | Advanced/ innovative | | Substantial, good info | |
| | n/N | % | n/N | % | n/N | % | n/N | % | n/N | % |
| Quality | 7/10 | 70 | 0/10 | 0 | 8/10 | 80 | 3/10 | 30 | 5/10 | 50 |
| Competition | | | 0/10 | 0 | 6/10 | 60 | 3/10 | 30 | | |
| Safety integral to business success | 11/21 | 52 | 0/21 | 0 | 14/21 | 67 | 7/21 | 33 | 9/21 | 43 |
| Whole sample | 20/66 | 30 | 9/66 | 14 | 31/66 | 47 | 11/66 | 17 | 16/66 | 24 |
| Functionality | | | | | 9/33 | 27 | | | | |

*Note:* In Table 8.2, N is the number of firms in which a particular motivation applied, and n is the number of those firms performing at the stated level for the particular substantive safety outcome.

outcomes. As set out in Table 8.2 above, these 21 firms were more likely to have comprehensive hazard recognition (52 per cent (11/21)) and none were blinkered (0 per cent (0/21)). They were also more likely to use safe place controls as the primary risk control measures (67 per cent (14/21)), to use some advanced or innovative controls (33 per cent (7/21)), and to provide substantial, good quality safety information (43 per cent (9/21)).

Two economic motivations, cost and reputation, did not differentiate manufacturer performance for substantive safety outcomes. The firms motivated by these concerns performed similarly to firms in the sample overall (hence these motivations are not included in Table 8.2).

On the other hand, manufacturers motivated by concerns about machinery functionality performed more poorly compared with the sample overall, specifically for risk control. These firms considered that particular risk control measures, or increasing the level of risk control, would impede the functionality of their machinery. As set out in Table 8.2 above, compared with the sample overall, firms motivated by functionality concerns were less likely to use safe place controls as the primary risk control measures (27 per cent (9/33)). There was no evidence that they compensated for poor risk control by providing good quality safety information as their performance for information provision was similar to the sample overall.

The eight manufacturers whose action on machinery safety matters was driven by those they did business with as customers or distributors tended to have poorer performance for all substantive safety outcomes. (This trend is not presented in Table 8.2 due to the smaller number of firms motivated in this way). The perception in these firms that their customers or distributors were taking responsibility for machinery safety, and did not require them to make decisions about safety, provided them with justifications for not proactively addressing safety issues in machinery design and construction. They variously believed that their customers or distributors would advise them if risk control measures were required or would incorporate them themselves, or that the customer or distributor would produce the machinery safety information.

The nine firms concerned that there was an uneven playing field in the market for their machinery, because of different expectations by OHS regulators, tended to have poorer performance for risk control in particular. They all relied on safe person measures for some risks and, where they had developed more advanced or innovative solutions for some risks, they sold these as optional models to avoid an adverse impact on sales, or did not implement a safe design solution unless they had established that it was acceptable to the market in regard to cost and functionality. Nor was there any evidence that these firms compensated for their reliance on safe person measures by providing good quality safety information (again due to smaller numbers these trends are not included in Table 8.2 above).

*Values, Attitudes and Substantive Outcomes*

Of the values and attitudes mediating manufacturers' responses on machinery safety matters, the only one with the potential to support good performance for substantive safety outcomes was the moral obligation to protect human health and safety. Consistent with this, the nine firms in which key individuals expressed this value, tended to have better performance across all safety outcomes. However, a sense of moral obligation did not guarantee good performance by all of these firms because, as discussed above, the firms in which key individuals espoused this value were also occupied with commercial goals and the normative motivation had to compete with economic motivations.

The unsafe worker attitude impeded performance for substantive safety outcomes, particularly for risk control, as this attitude provided key individuals with justifications for not redesigning their machinery, or not incorporating risk control measures in machinery which they knew to be hazardous. As set out in Table 8.3 below, compared with the sample overall, firms in which key individuals expressed this attitude were less likely to use safe place controls as the primary risk control measures (33 per cent (12/36)). The greater reliance on safe person measures in these firms was consistent with the attitude that users should act more safely. This did not, however, encourage these firms to provide better machinery safety information to promote safe behaviour and few provided substantial, good quality information (11 per cent (4/36)). A key reason was the view of key individuals in these firms that customers and end users do not read or apply machinery safety information. This perception co-existed with the unsafe worker attitude in 15 of these firms.

**Table 8.3    The unsafe worker attitude and substantive safety outcomes**

| | Substantive safety outcomes | | | | | | | | |
|---|---|---|---|---|---|---|---|---|---|
| | Comprehensive | | Blinkered | | Safe place emphasis | | Advanced/ innovative | | Substantial, good info | |
| | n/N | % | n/N | % | n/N | % | n/N | % | n/N | % |
| Whole sample | 20/66 | 30 | 9/66 | 14 | 31/66 | 47 | 11/66 | 17 | 16/66 | 24 |
| Unsafe worker | | | | | 12/36 | 33% | | | 4/36 | 11 |

*Note:* In Table 8.3, N is the number of firms in which the attitude applied, and n is the number of those firms performing at the stated level for the particular substantive safety outcome.

Finally, and unsurprisingly, the eight firms in which key individuals expressed the attitude that safety has gone too far tended to have poor performance across all substantive safety outcomes (again, due to smaller numbers this trend is not presented in Table 8.3).

*Summary of Motivational Factors and Substantive Outcomes*

Manufacturers responded to motivational factors, which differed in the context of each firm's operations and interactions with external actors, and the values and attitudes of key individuals in firms. Some firms regarded machinery safety as integral to the success of the business, or at least a key commercial goal. These were the firms motivated to manage business risks or quality, to realize business opportunities, or to secure a competitive advantage or ensure competitiveness. Some firms accepted Australian OHS law, the European regulatory regime for machinery safety or technical standards as legal or quasi-legal imperatives for addressing machinery safety matters, or recognized the potential benefits of being able to market their machinery as complying with particular legal requirements or technical standards. These firms with positive economic, legal or quasi-legal rationales for addressing safety in machinery design and construction tended to perform better on at least some safety outcomes. In some of these firms, the moral obligation of key individuals to protect human health and safety reinforced legal, quasi-legal or economic motivations for addressing safety matters.

The other legal imperatives, inspection and enforcement by Australian OHS regulators and concern about legal liability, were generally neither sufficient to drive firms to inform themselves about the type of action needed, nor to encourage them to take sustained action to improve their performance for substantive safety outcomes. These legal motivations, as well as several economic motivations and the negative attitudes of some key individuals, were consistent with firms not accepting and taking responsibility for machinery. In particular, firms whose responses were shaped by their customers or distributors, by concerns that the functionality or competitiveness of their machinery might be impeded by safety measures, and/or by negative attitudes about unsafe workers or that safety has gone too far, had justifications for not taking the type of action necessary to improve machinery safety. These firms performed poorly on some or all of the substantive safety outcomes.

The findings presented here indicate that it is not sufficient for research to document motivations claimed by regulatees without examining what, if any, action or outcomes are supported by particular motivations. In the present research, for example, claimed reputational concerns were not linked with better performance for substantive safety outcomes.

The research with machinery manufacturers has also indicated that specific motivational factors did not, by themselves, explain firms' performance for the substantive safety outcomes. In particular, the factors linked with better performance did not ensure better performance in all firms. Rather, the combination of and interplay between motivational factors in the context of each firm's operations, together with the knowledge of those involved in machinery design and construction, determined whether the action taken by the firm was sufficient to support good performance for substantive safety outcomes. This issue is taken up in Chapter 9, which presents a theoretical framework to explain how knowledge and motivational factors combined to shape a particular firm's response on machinery safety matters, and the firm's performance for substantive safety outcomes.

## Conclusion

This chapter has demonstrated that machinery manufacturers were motivated by their experience or perception of legal pressures, their commercial goals to ensure the marketability of their machinery and firm profitability, and the values and attitudes of key individuals in firms. These motivational factors provided firms with rationales for taking action on machinery safety matters, or justifications for limiting or not taking such action. In essence, firms' construction of reality either supported or impeded action on machinery safety.

The findings presented here extend explanations in the safety and socio-legal literatures of the contextualized nature of organizations' responses to regulation and the role of motivational factors. They have demonstrated the plurality of motivational factors in relation to safety in machinery design and construction. The research has located state regulation (the law and its inspection and enforcement) among these motivational factors but as weaker influences, which must compete with other performance shaping factors. The research has also provided a nuanced account of co-existing motivational factors in firms, revealing that motivational factors may be inconsistent in their influence and pull firms' safety effort in different directions (see also Kagan, Gunningham and Thornton, 2011).

By examining manufacturers' performance for substantive safety outcomes, as distinct from their means or processes for complying, the research has provided additional insights into different types of motivational factors. It has shown that manufacturers that accepted responsibility for machinery safety and had positive legal, quasi-legal or economic rationales for taking action on safety matters tended to perform better for substantive safety outcomes, compared with the sample overall. On the other hand, firms that did not accept responsibility for machinery safety tended to perform more poorly. This included firms that only took action when prompted to do so by inspection and enforcement by OHS regulators, or directions from their customers and distributors, firms more concerned about machinery functionality and competitive disadvantage, and firms in which key individuals had negative attitudes about end users or safety generally.

The evidence in this chapter about the influence of customers and distributors reinforces other findings in this research, which suggest that a less optimistic view is warranted in regard to the potential for non-state actors to positively influence the performance of firms with which they have commercial relationships. The next chapter establishes the significance of these findings in clarifying the nature and extent of non-state influences on manufacturers' performance for machinery safety.

For motivational factors generally, the findings in this chapter suggest that regulators and policy makers, professional and industry stakeholders who seek to improve manufacturers' responses on machinery safety, will need to take account of the different motivational factors in the context of particular firms' operations and their interactions with external actors. As Haines (1997, pp. 215, 224) has argued, greater attention needs to be paid to the dynamics within and outside firms that influence their behaviour. The separate legal, economic and individual strands of motivational factors can provide a useful analytical framework for identifying the points of greater leverage over firms' performance (see also Gunningham, Kagan and Thornton, 2003, p. 149).

This research makes a further contribution in confirming that both motivational factors and knowledge shaped firms' responses on machinery safety. The next and final chapter crystallizes this finding, theorizes responses to safety in machinery design and construction, and discusses the implications of the research for regulators and policy makers, and professional and industry stakeholders.

# Chapter 9
# Conclusions, Theory and Policy Implications

If manufacturers do not produce machinery that is inherently safe the consequences are grave, and measured in fatalities, injuries and illness (AWCBC, 2012; Driscoll, et al., 2008; European Commission, 2008; Fitzharris, et al., 2011; Harris and Current, 2012; Safe Work Australia, 2009; 2011; 2013a). With this in mind, the empirical evidence from this research underscores the need to lay down firm foundations for both regulatory policy and industry practice for safety in machinery design and construction.

The research has distinguished three levels of performance among manufacturers (see Chapter 3). Exceptional performers substantively complied with the regulatory goal of preventing death, injury and illness, as they comprehensively recognized the different types and instances of hazards for their machinery and incorporated more effective safe place[1] controls as the primary risk control measures. They also provided substantial safety information in a combination of labels, manuals or other forms, and which was easy to locate, read and understand.

A second group of manufacturers were mediocre performers who had not recognized some types or instances of hazards for their machinery, relied on safe person measures that required end users to actively avoid risks and take care to protect themselves and others, and/or provided safety information that was limited in scope or hard to locate, read and understand. A third group were poor performers as they only recognized mechanical hazards for their machinery, relied on safe person measures for some risks and provided very little, or poor quality safety information, or none at all. Neither the mediocre nor the poor performers substantively complied with the regulatory goal of prevention.

The research has also demonstrated manufacturers' rather mixed responses for assessing and managing risks (see Chapter 7). Only four firms, all large, had systematic arrangements for addressing safety in machinery design and construction, consistent with the wider trend for management systems to be confined to large or capital intensive organizations (Hale, 2003; Lamm and Walters, 2004; Parker, 2002, pp. 56–7). A wider group of 39 firms conducted some form of risk assessment, or engaged a consultant to do this, but the quality and rigour of assessments differed markedly, ranging from the ritualistic generation of paperwork to a thorough and informed investigation of hazards, risks and preventive measures from the design stage. Such disparity in the efficacy of risk

---

1  On the distinction between 'safe place' and 'safe person' control measures see Atherley (1975; 1978) and Haddon (1973; 1974; 1980).

assessment resonates with European experience (Boy and Limou, 2003, pp. 37–9, 62–3; Crabb, 2000, pp. 20, 39; Jensen, 2001; 2002b; Karageorgiou 2000, p. 284).

As important is the contribution this research makes to explaining manufacturers' practices and performance for substantive safety outcomes. This is the focus of the final chapter of the book, which draws together the empirical, conceptual and theoretical contributions of the research, and sets out some policy implications and strategic directions.

The first step is to consolidate the empirical evidence about the construction of knowledge, the nature and influence of motivational factors (motivations, values and attitudes), and how these respectively shaped firms' actions and performance. Next the evidence about the role and influence of state regulation and non-state actors is drawn together. The chapter then turns to theorizing the relationship between knowledge, motivational factors, actions, performance and compliance with regulatory goals. The final contribution is the normative one. The policy implications of the research are outlined, and some strategic directions are proposed for state regulation of machinery safety and for building the capacity of manufacturers to perform well for substantive safety outcomes.

## The Construction of Knowledge through Practice

Safety and socio-legal scholarship have recognized the central role of capacity, including knowledge and skills, in business self-regulation and attention to safety (Hale and Hovden, 1998; Hutter, 2001, pp. 301–2; Nytrö, Saksvik and Torvatn, 1998; Parker 2002, pp. ix–x, 57; Parker and Nielsen 2011, p. 5, 15–17). There is also a view that firms may learn how to comply with legal obligations through their business relationships with customers, suppliers and other enterprises (Gunningham and Sinclair, 2002, pp. 17–18; Lamm and Walters, 2004; Walters, 2001, pp. 52, 375). For designers, European researchers have documented their learning through experience and interactions with others, their preference for technical standards over other sources, and their practice of drawing on suppliers for information (Broberg, 1997; 2007; Crabb, 2000, pp. 26–7; Hale and Swuste, 1997; Swuste, Hale and Zimmerman, 1997). The present research advances these insights by highlighting the principal and minor constituents of manufacturers' knowledge about safety, the contribution of both social and individual processes to learning, and how multiple bases for learning are part of the explanation for differences in firm performance.

The research has established that those involved in machinery design and construction learned about machinery safety matters through these activities, and has confirmed three aspects of practice as the principal constituents of knowledge – drawing upon experience producing machinery in their own and other firms, interacting with customers and referring to technical standards. The research has also revealed the highly situated nature of learning in the context of a particular firm's operations and interactions with external actors. Within and additional to the three central practices, situational manifestations of practice gave rise to multiple bases for construction of knowledge about safety matters, and different perspectives on safety problems, ways to resolve them, and approaches to assessing and managing risks. These multiple bases included Australian, European and international technical standards, component suppliers, customers and end users, and information about injuries, incidents or hazardous exposures involving machinery.

Crucially, the constituents of knowledge went beyond authoritative sources such as the legal obligations for machinery safety, occupational health and safety (OHS) regulators' guidance and the considerable specialist body of knowledge. These authoritative sources were less common constituents of knowledge in only a small proportion of firms. There were few examples of firms learning from legal instruments or OHS regulators (see 'The Influence of State Regulation' below). Also, with only two firms making use of ergonomics references, those involved in machinery design and construction generally did not have the benefit of specialist insights into designing inherently safe machinery, integrating safety and functionality, minimizing the risks of unintended use and producing useful safety information.

As well as establishing the constituents of knowledge about machinery safety in manufacturing firms, and the diversity of these, the research has determined that individual processes also came into play in learning about safety matters, consistent with a social constructivist perspective of learning (Billett, 2001; 2006; Palincsar, 1998; and see Chapter 6). Those involved in machinery design and construction, as key decision makers and designers, had different foundations for learning due to their disparate professional and vocational backgrounds as engineers, other professionals and tradespeople of different types, and end users of machinery. As a consequence, individual learning about machinery safety and the collective know-how in firms differed according to both the constituents of knowledge, and how these were interpreted and laid down as knowledge by individuals with diverse personal histories and capacities.

In turn, the varied constituents and foundations for learning about safety differentiated how, and how well, manufacturers addressed safety in machinery design and construction. Certain bases for constructing knowledge were linked with better performance for some or all of the substantive safety outcomes. These were: input from end users at the design stage; specialist human factors, safety engineering or general safety resources; information about injuries, incidents and the end use work environment; assistance from suppliers; and information from trials of prototypes or models. On the other hand, firms performed more poorly when they relied on customers to tell them what to do, or received no input from customers or end users at all. There was also some evidence to suggest that engineers had more skills for gathering, interpreting and applying information, and a more relevant compendium of knowledge to support better performance for machinery safety, compared with tradespeople or individuals with no qualification or trade.

The common practice of referring to technical standards did not set firms apart from those that did not refer to these standards. It is likely that the ad hoc use of technical standards by those involved in machinery design and construction rather than more rigorous application of standards, coupled with weaknesses in particular standards, contributed to the unexceptional performance of firms that referred to technical standards (see also Backstrom and Döös, 2000; Boy and Limou, 2003, pp. 55–8, 88–9, 120; KAN, 2008; Worringham, 2004).

In summary, in digging deeper into the conceptual theme of capacity this research has revealed that diverse practices and individual factors differentially shaped learning and, in turn, manufacturers' performance for substantive safety outcomes. Processes for learning did not, however, completely explain firm performance, which was also shaped by motivational factors in the form of business motivations, and the values and attitudes of key individuals in firms. These, like knowledge, were contextualized in the operations of particular firms and their interactions with external actors.


**Motivational Factors**

Safety and socio-legal scholarship has offered contextualized explanations of firms' or individuals' responses to regulation with reference to plural legal, economic, social or normative motivations, and values and attitudes (Braithwaite V, 1995; Braithwaite V, et al., 1994; Braithwaite, Murphy and Reinhart, 2007; Genn, 1993; Gunningham, Kagan and Thornton, 2003, chs. 3, 6; 2005; Haines, 1997, ch. 7; Hopkins 1995, ch. 11; Kagan, Gunningham and Thornton, 2011; May, 2005; Parker, 2002, ch. 3). Among study firms economic motivations derived from their commercial goals of ensuring the marketability of their machinery and firm profitability. They concerned risk minimization, reputation, business opportunities, quality, functionality, cost, competition, and relationships with customers or distributors. Legal and quasi-legal motivations derived from the perceived authority of the law or particular instruments, and the threat of penalties. They related to Australian OHS law, inspection and enforcement by OHS regulators, the European regulatory regime for machinery safety, a particular technical standard (Australian, international or European), or a non-specific concern about litigation or legal liability.

These various economic, legal and quasi-legal motivations were the factors driving or constraining firms' action on machinery safety. They were supplemented by the values and attitudes of key individuals which provided assumptions about responsibility for safety, and justifications for taking, or not taking, particular action.

Machinery manufacturers yielded rich data, not only about the plurality of motivational factors, but also about the interplay between co-existing and sometimes conflicting factors. The mix of motivations, values and attitudes in a particular firm might provide mutually reinforcing rationales for taking action on machinery safety, mutually reinforcing justifications for limiting or not taking such action, or conflicting pressures drawing firms' decision making and action in different directions.

The research has also shed light on whether particular motivational factors differentiated firm performance and, if so, how. Manufacturers that accepted responsibility for machinery safety and had positive rationales for taking preventive action tended to perform better for substantive safety outcomes (compared with the sample of firms overall). In particular, firms that accepted Australian OHS law, the European regulatory regime for machinery safety or particular technical standards, as legal or quasi-legal imperatives for taking action on machinery safety, tended to perform better. This was also the case for firms that identified machinery safety as being integral to the success of their businesses, or at least integral to their economic motivations to manage business risks or machinery quality, or to secure business opportunities or a competitive advantage. In some of the firms with positive rationales for taking preventive action, key individuals also expressed a moral obligation to protect human health and safety and this value reinforced legal, quasi-legal or economic motivations for taking action on machinery safety.

On the other hand, firms that did not accept and take responsibility for machinery safety tended to perform poorly for some or all of the substantive safety outcomes. This was the case for firms motivated by 'after the event' inspection and enforcement by OHS regulators following an incident or complaint, and the non-specific concern about legal liability. These motivations were generally neither sufficient to drive manufacturers to inform themselves about the type of preventive action needed, nor to encourage them to take sustained action to improve their performance for machinery safety. In addition, firms driven by their customers or distributors, or concerns about machinery functionality or reduced competitiveness, and firms in which key individuals had negative attitudes about unsafe workers or that safety has gone too far, had justifications for not taking action on machinery safety.

The research has also highlighted the significance of the unsafe worker attitude in manufacturing firms, a perspective attributing work-related incidents to worker careless or unsafe behaviour (Dejoy, 1994; Glendon, Clarke and McKenna, 2006, pp. 83–91). In the context of manufacturers' commercial goals relating to machinery functionality, cost and competitiveness, the unsafe worker attitude provided a justification for firms to continue to produce and supply machinery that they knew to be hazardous, pointing to the need to challenge this attitude in order to foster effective risk control action (Hasle, Kines and Andersen, 2009).

In summary, this account of motivational factors advances understanding of the particular goals and priorities of machinery manufacturers but, taking a wider view of business behaviour, it illustrates the complexity of motivational factors. Understanding the plurality and concurrence of motivational factors, and the potential for conflict and reinforcement among them, are essential to explaining and, when necessary, challenging their impact on firm performance.

We have also seen that both state regulation and non-state actors are part of the mix of motivational factors, as well as being constituents of knowledge about machinery safety matters. What then can we make of the extent and nature of their respective influences? This is the focus of the next two sections.

## The Influence of State Regulation

Australian OHS regulators sought to influence manufacturers' attention to safety in the design and construction of machinery through OHS statutes, regulations and approved codes of practice coupled with some, albeit limited, compliance support, inspection and enforcement activities (Johnstone, 1997, pp. 261–3; Johnstone, Bluff and Clayton, 2012, pp. 331–5; and see Chapters 2 and 5). Firms supplying into overseas markets might also come across state regulation in any of those markets, although the globally influential European regulatory regime implementing the *Machinery Directive* was the only one that some study firms were aware of (European Commission, 1998a; 1998b; 1998c, 2006; European Standards Organizations, 2009; IMS Research, 2009; and see Chapters 2 and 4).

In principle, these Australian and European regulatory regimes might contribute to the knowledge base and motivation of firms to address safety in machinery design and construction but, in practice, their influence in study firms was modest at best as many were not aware of the regulatory systems' expectations of them. There were only three firms in which key individuals, aided by in-house safety advisers, had more detailed knowledge of Australian OHS law to inform their decision making. This body of law was a minor constituent of the knowledge base in a further 22 firms, which had picked up some information about the law through interactions with their customers, distributors or other industry sources (see Chapter 4). There were also 11 firms in which the key individuals were aware of the European regulatory regime but only three of these firms, all of which conducted their own conformity assessment, had staff who knew some specific details about the European requirements.

Low awareness of relevant legal obligations could not be attributed solely to properties[2] of the law, a factor identified in some previous studies of business responses to safety-related law (Cowley, Culvenor and Knowles, 2000; Crabb, 2000, p. 26; Fairman and Yapp, 2005a; Genn, 1993; Hutter, 2001, pp. 86–7, 96). In study firms, there was a more fundamental problem of not attempting to engage with legal obligations or authoritative sources of guidance, which had to take their place among multiple, everyday sources of information embedded in the operations of firms and their interactions with external actors (see Chapter 4). It would take more than carefully crafted legal obligations to ensure that firms engaged with their obligations.

Nonetheless, even when key individuals did not learn about their legal obligations accurately and comprehensively, the perceived authority of the law

---

2  On properties of the law see Baldwin (1995, pp. 9, 11), Black (1997, pp. 22–3), Diver (1983) and Scott (2010, pp. 108–11).

contributed to the motivation in some firms to conduct an assessment of their machinery, engage a consultant to do this, or take some other form of preventive action (14 firms for Australian OHS law, and 11 firms for the European regulatory regime). Such a motivation did not, however, necessarily prompt firms to pay attention to whether the action they took and the outcomes they achieved complied with their legal obligations (Australian or European), and these obligations had to take their place among other motivations and attitudes which, in some firms, conflicted with any positive motivation engendered by legal obligations (see Chapter 8).

Compliance support, inspection and enforcement activities of Australian OHS regulators were a weak influence, as inspectors had discretion (see Black, 2001b) to choose whether and how to interact with firms as designers or manufacturers, which they mostly did as event-based enforcement in response to incidents or complaints (Johnstone 2003, chs. 3–5; and see Chapter 5). Non-compliance was treated as a one-off issue rather than regulators attempting to secure commitment, capacity and arrangements to self-regulate and address safety on an ongoing basis (Johnstone, 2004b, p. 147–8; Johnstone and Jones, 2006; Parker, 2002, pp. ix–x, 27, 43–61). Matters of principle such as consistency and fair treatment (Murphy, 2005; Tyler, 1997; Yeung, 2004, pp. 36–43) also warranted more attention as inspectors were inclined to treat similar problems differently, when they sought particular action by one firm and different action, or no action, by the firm's competitors.

A leading regulatory theory, responsive regulation, recommends that regulators engage strategically with regulatees by applying a hierarchy of mechanisms. The approach begins with cooperative dialogue to pave the way for productive problem solving and escalates enforcement, through mechanisms such as notices and higher order sanctions, if non-compliance continues (Ayres and Braithwaite, 1992, pp. 21–2, 35–41; Black and Baldwin, 2010; Braithwaite, 2011; Johnstone 2004b, pp. 155–60). In their dealings with machinery manufacturers, inspectors did not practice responsive enforcement. They favoured a cooperative approach, but principally used persuasion without providing advice, rarely issuing notices and prosecution was exceptional.

From the perspective of study firms, OHS regulators were not key constituents of knowledge about machinery safety, with only three firms learning something through OHS regulator advice, guidance material or, in one case, a prosecution transcript (see Chapter 5). In this respect, the research supports a growing number of studies showing that regulatees' uptake and use of guidance materials provided by OHS regulators is low and that businesses, especially smaller ones, have a preference for oral advice (James, et al., 2004; Mayhew, et al., 1997; Melrose, et al., 2006; Wiseman and Gilbert, 2002; Wright, Marsden and Antonelli, 2004, pp. i, vii; Maxwell, 2004, pp. 259, 262–5).

Inspection and enforcement by OHS regulators had a wider influence as a motivation, although its impact on manufacturer performance was limited. Twenty firms were motivated to take some action as a result of inspection and enforcement experienced directly, or awareness of inspection and enforcement through their customers or distributors. Their actions included conducting risk assessment, controlling particular risks, redesigning machinery and providing safety information (see Chapter 5). These firms did not, however, perform markedly differently from the sample of firms overall, for any of the substantive safety outcomes (see Chapter 8). Inspection and enforcement activities were capable of capturing firms' attention, but were insufficient to prompt sustained action and ensure better performance for substantive outcomes.

The findings for general deterrence arising from Australian prosecutions of machinery designers or manufacturers (or suppliers), are consistent with socio-legal scholarship on general deterrence in different regulatory regimes (Fairman and Yapp, 2005a; Gray and Scholz, 1990; Gunningham, Thornton and Kagan, 2005; Jamieson, et al., 2010; Thornton, Gunningham and Kagan, 2005). Most firms were not aware of such enforcement because it was too infrequent, and those that were aware of relevant cases typically heard about them second or third hand, and did not have sufficiently reliable information to know what action to take.

In summary, state regulation in the form of the Australian and European legal obligations for machinery design and construction, and the compliance support, inspection and enforcement activities of OHS regulators, must be located as minor constituents of knowledge and as weak motivational strands among a wider web of influences shaping firms' responses on machinery safety. The research points to the need for state regulatory interventions to ensure that firms engage with, understand and implement their legal obligations, and suggests that this will require carefully crafted strategies to build capacity, and inspect and enforce compliance, as discussed below (see 'Policy Implications and Strategic Directions').

## Regulatory Influences Beyond the State

In regulatory theory, the concepts of regulatory space and de-centred regulation recognize the role of regulatory actors beyond the state agencies that set, administer and enforce legal obligations (Black, 2001a; Hancher and Moran, 1989; and see also Baldwin, Scott and Hood, 1998, pp. 3–4; Hutter, 2006, 2011). For machinery manufacturers these non-state actors included their customers, distributors or suppliers, industry associations, and professional advisory services and consultancies. Technical standards bodies[3] also fit into this category, although some of the standards they provide may be adopted under state regulatory regimes, as with some Australian standards and European harmonized standards (see Chapter 2).

Perspectives on the regulatory role of non-state actors in safety and socio-legal scholarship are mixed. Some scholars suggest there is the potential for large customers, suppliers, professional advisers such as accountants, and industry

---

3    For example the European Committee for Standardization (CEN), the International Organization for Standardization and Standards Australia.

associations to contribute positively to the know-how and capability of the firms with which they have business relationships, by communicating regulatory messages to them, or checking and influencing their compliance with state regulation (Gunningham and Johnstone, 1999, p. 133; Gunningham and Sinclair, 2002, pp. 17–18; Hopkins and Hogan, 1998; Lamm and Walters, 2004; Walters, 2001, pp. 52, 375–6; Walters, 2002, pp. 45–6). However, some empirical studies of firms' responses to safety-related regulation suggest that the influence of non-state actors is weak in practice (Gunningham, Thornton and Kagan, 2005; Hasle, Bager and Granerud, 2010; Hutter, 2011, ch. 5; Hutter and Jones, 2007; James, et al., 2004, pp. 100–1). Limiting factors include the knowledge or resource constraints of particular non-state actors, the less than robust information provided by these sources and regulatees not knowing what action they should take.

The research with machinery manufacturers lends support to both perspectives. It confirms that certain non-state actors contributed to the knowledge and motivation in firms to address safety matters, but reinforces reservations about the nature and extent of that influence and the potential to foster compliance with regulatory goals through these sources.

Interactions with customers, other producers of machinery, component suppliers and industry sources more widely were key bases from which those involved in design and construction activities learned about state regulation of machinery safety, safety problems and solutions to them, and methods for risk assessment (see Chapters 4, 6 and 7). Indeed, as interactions with these non-state actors were central to design and construction practice, it was axiomatic that those involved in these activities would learn through these sources. Large customers or distributors also contributed to the motivation to take action on machinery safety matters in some firms, and some firms were motivated by messages about inspection and enforcement relayed by their customers or distributors (see Chapters 5 and 8).

Technical standards were a key information source whether these were Australian, international (ISO) or European harmonized standards, or sourced from other standards bodies. Particular technical standards were also a driving force for action on machinery safety in some firms, because of the perceived authority and commercial value of being seen to comply with the standards, and whether or not the particular standard had any legal standing (see Chapter 8 and Productivity Commission, 2006, pp. 6, 14).

Beyond confirming that various non-state actors contributed to the knowledge base or motivation to address machinery safety in some firms, the research has revealed that knowledge and motivations constituted through non-state sources did not, in general, equip firms to perform well for substantive safety outcomes, or to conduct timely, logical and thorough risk assessments. The two exceptions were assistance from component suppliers, which was linked with better performance for risk control, and particular technical standards that were a driving force for action on machinery safety (not just a source of information).

For the study firms overall there was little evidence that, through their interactions with non-state actors, they had internalized their legal obligations for the safe design and construction of machinery, or the regulatory goal of preventing death, injury and illness. Interacting with customers (or distributors) was widespread, but the only link with better performance was for manufacturers that managed the interaction and actively sought input from end users in customer firms in the course of designing machinery (see Chapter 6). On the other hand a group of smaller manufacturers, whose action on machinery safety was driven by large customers, did not take sustained action to address safety and performed poorly for all substantive safety outcomes. The large customers took charge of how the machinery was designed and constructed, set the standards to be met, and the small manufacturers only took action on machinery safety matters within the limits of particular customer/distributor requirements (see Chapter 8). Also, individuals who knew of key cases (*Arbor Products*[4] and *Viticulture Technologies*[5]), through industry associations or other industry sources, did not know what action they should take as they either had insufficient information or were misinformed about the cases through these sources. The advice they received reinforced the negative attitude, that end users of machinery should take more responsibility for safety, rather than prompting constructive preventive action (see Chapter 5).

This research signals the need for a nuanced exposition of the influence of non-state actors in regulatory processes, and in regulatory theory and models explaining business conduct. While non-state actors play a role, their influence may run counter to regulatory goals. Regulators and others seeking to build firms' capacity or motivations to comply will need to monitor and mediate the influence of non-state actors, and not simply harness or rely on them to constructively influence firm performance.

## Explaining Manufacturers' Responses on Machinery Safety

In addition to providing empirical evidence of manufacturers' performance and the principal elements shaping their performance, this research makes conceptual and theoretical contributions. Through a recursive process of inductive reasoning from the empirical data for the 66 study firms and deductive reasoning from the literature, substantive theory was constructed in the form of an integrative interpretation of the patterns and relationships in the empirical data (Marshall and Rossman, 2006, pp. 161–2; Mason, 1996, p. 142; Morse and Richards, 2002, pp. 169–70; Neuman, 1997, pp.46–8, 55–6; Richards, 2005, pp. 128–34; Silverman, 2001, pp. 237–40). This explained the relationship between knowledge and motivational factors in manufacturing firms, their action and performance for substantive safety outcomes and compliance, or non-compliance, with the regulatory goal

---

4 *WorkCover Authority of New South Wales (Inspector Mulder) v Arbor Products International (Australia) Pty Ltd* (2001) 105 IR 81.

5 *Shepherd v Viticulture Technologies (Aust) Pty Ltd* (unreported, Court of Petty Sessions, Albany (WA), Malone SM, charge no 1941/01, 15 May 2003).

of prevention. This theorizing links to social constructivist learning theory (Billett, 2001; 2006; Palincsar, 1998) and bounded rationality theory (Gigerenzer and Selten, 2001; Simon, 1955), and extends understanding about the interplay between commitment, capacity and arrangements as essential ingredients for self-regulation (Johnstone and Jones, 2006, pp. 485–6; Parker, 2002, pp. ix–x, 43–61; Parker and Nielsen, 2011, p. 5).

The inductive reasoning was grounded in analysis of: the characteristics and attributes[6] of study firms, and their key individuals and designers; firms' actions and practices for machinery safety; the state regulatory regimes and non-state actors influencing firms' responses; and understandings of machinery safety matters among key individuals in firms (see also the Appendix). Health and safety, socio-legal and learning literatures were applied in comparing and contrasting explanation grounded in the data with pre-existing knowledge and theory (Flick, 2006, p. 59; Morse and Richards, 2002, pp. 169–70). Also grounded in the empirical data and defined with reference to the literature are nine central concepts relating to knowledge, motivational factors and firms' responses, which are the building blocks of this theorizing (Neuman 1997, pp. 39–41). These concepts were introduced in earlier chapters. They are represented in Figure 9.1 below and defined in the following explanation to ensure that each concept can be readily understood, operationalized and verified (Glaser and Strauss, 1967, p. 3).

Applying social constructivist learning theory (Billett, 2001; 2006; Palincsar 1999), knowledge about machinery safety is conceived as the product of the constituents through or from which individuals construct knowledge, principally through practice, and the individual factors which frame their interpretation of these constituents and their construction of knowledge. The three conceptual building blocks here are machinery safety knowledge, constituents of such knowledge and individual factors, which are defined as follows (see also Chapter 6).

The concept of *machinery safety knowledge* encompasses all that a person knows or believes to be true about machinery safety, including information, skills, experiences, beliefs and memories about machinery safety matters. This includes factual information, knowledge of processes and how to use them, and discourse knowledge about language and its use, among other forms (Alexander, 1991; Billett, 2001). *Constituents of knowledge* are the experiences, interactions with others, information materials and sources, and situational manifestations of practice in firms through and from which individuals construct knowledge about machinery safety. These constituents are wider than, and may or may not include, authoritative sources such as legal instruments, regulator advice or guidance materials, or specialist resources dealing with safe design and construction. *Individual factors* are individuals' professional qualifications, vocational or trade backgrounds, work experience and other aspects of their personal histories and

---

6  Firm characteristics and attributes included size and location, type of machinery produced and markets. Individual attributes included key individuals' and designers' qualifications and experience.

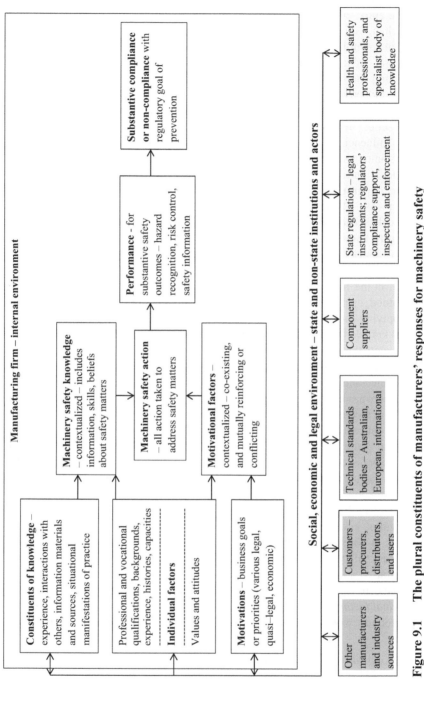

**Figure 9.1    The plural constituents of manufacturers' responses for machinery safety**

capacities. They constitute the interpretive frame through which individuals interpret and construct knowledge from the constituents of knowledge that they encounter. In the context of this research individual factors concern the key individuals (decision makers) and designers as the principal actors in machinery design and construction.

Together with individual knowledge, and collective knowledge in firms, motivational factors shape firms' action on machinery safety, with both elements having the potential to support or constrain such action. The three concepts for motivational factors are motivations, values and attitudes, as defined below (see also Chapter 8).

*Motivations* are the reasons behind or factors driving an individual or firm's actions (Glendon, Clarke and McKenna, 2006, pp. 104–5, 108; Reber and Reber, 2001, pp. 447–8; Sundström-Frisk, 1996). They are the perceived inducements or rewards for taking or not taking particular action. *Values* are deep-seated principles or standards of behaviour that constitute motivational goals and guide the way individuals evaluate situations, select actions, and explain or justify their actions (Glendon, Clarke and McKenna, 2006, pp. 187, 189; Licht, 2008, pp. 3, 21–2, 41; Reber and Reber, 2001, p. 783; Sundström-Frisk, 1996; 1999). They reflect ideals or qualities that are desirable means or ends in themselves. *Attitudes* are settled ways of thinking and reflect an individual's learned tendencies to act in a consistent way towards a particular person, situation or thing. They can influence behaviour via motivation, reinforce individual desires and needs, and provide a defence mechanism to protect the individual from harsh realities (Glendon, Clarke and McKenna, 2006, pp. 187–9, 207; Reber and Reber, 2001, p. 63; Sundström-Frisk 1996; 1999).

By their nature, both motivational factors and knowledge are contextualized in the operations of firms and their interactions with external actors. State legal obligations for machinery safety (Australian and European), and regulators' compliance support, inspection and enforcement activities, contribute to knowledge and motivations in some firms but, even when they have some influence, they must compete with the other, non-state, constituents of machinery safety knowledge and motivations in firms.

In turn, the quality and rigour of firms' machinery safety actions shape their performance and substantive compliance, or non-compliance, with the regulatory goal of prevention. *Machinery safety action* includes all action taken by firms to address machinery safety, including assessing and managing risks, incorporating risk control measures in machinery, applying particular technical standards, producing or improving safety information, and any other preventive action (see Chapters 6 and 7). The concept of *performance* is the level of performance achieved by firms for substantive safety outcomes for hazard recognition, risk control and safety information, which are critical for preventing death, injury and illness arising from machinery (see Chapter 3). *Substantive compliance* is compliance with regulatory goals or objectives (Parker, 2002, p. 27), for example preventing death, injury and illness by ensuring that machinery is safe and without risks to health. In this research, the benchmark for substantive compliance is a firm comprehensively recognizing the hazards of the machinery, eliminating hazards or incorporating control measures which effectively minimize risks, and providing substantial, good quality safety information.

In some firms, knowledge, motivational factors and action translate into commitment, capacity and arrangements to achieve good performance for substantive safety outcomes, and comply with the regulatory goal of prevention. In other firms knowledge, motivational factors and action are insufficient to sustain compliance. An important implication of this theorizing is that machinery manufacturers make decisions within the limits of their contextualized knowledge, and steered by their motivations, values and attitudes. As a consequence, their decision making is often characterized by bounded rationality (Gigerenzer and Selten, 2001; Simon, 1955). Their imitation or copying of other firms' machinery or practices, making decisions that would achieve goals reflecting their motivations, choosing courses of action that appeared to them to be sufficient even if they were not the best course of action, were all features of bounded rationality (Gigerenzer and Selten, 2001; Klein, 2001; Selten, 2001; Simon, 1955). For example, bounded rationality was evident in the choice of firms to incorporate basic or industry standard control measures that were the product of imitation rather than innovation, and were compatible with machinery functionality but were not inherently safe.

Boundedly rational decision making is not sufficient to ensure compliance with either Australian or European legal obligations for machinery safety (see Chapter 2). To meet these standards manufacturers would need to be well informed about hazards, the adverse consequences of exposure to hazards and preventive measures to effectively control them. They would need, in essence, to make well-informed judgements rather than decisions characterized by bounded rationality.

This empirically grounded theorizing about manufacturers' responses for machinery safety builds on studies that have offered explanations of firms' or individuals' responses to regulation with reference to motivational factors (Braithwaite V 1995; 2009; Braithwaite V, et al., 1994; Braithwaite, Murphy and Reinhart, 2007; Hopkins, 1995; Thornton, Kagan and Gunningham, 2009), by highlighting the role of knowledge in shaping firms' responses. It extends scholarship establishing a role for both knowledge (capacity) and motivational factors (Genn, 1993; Gunningham, Kagan and Thornton, 2003; Kagan and Scholz, 1984; Parker, 2002; Parker and Nielsen, 2011), by providing a detailed explanation of the contextualized nature of knowledge and motivational factors, and the multiple, competing strands to these.

This theorizing complements Parker's (2002, ch. 2) conception of effective self-regulation by locating the specialist knowledge and commitment supporting self-regulation within the contextualized knowledge and wider motivations of firms. In so doing it offers a basis for analysing and understanding why firms perform at different levels, recognizing that those with mediocre or poor performance are unlike effectively self-regulating firms. This theorizing also takes a lens to the conceptual themes represented in Parker and Nielsen's (2011, pp. 5, 26) holistic and plural model of business compliance, reinforcing the conclusion that in order to explain business conduct there is a range of dynamic and interacting factors to take into account.

The findings of this research, and the theorizing about knowledge and motivational factors, have implications for state regulation, and for building business capacity to effectively address safety in machinery design and construction. The next section discusses these implications for policy and proposes some strategic directions.

## Policy Implications and Strategic Directions

*Implications for State Regulation of Machinery Safety*

Following a major national review into OHS laws in Australia, the Commonwealth and most state and territory governments have enacted harmonized legislation (Australian Government, 2008; 2009; Bluff and Gunningham, 2012; Safe Work Australia, 2013b). As a result, across much of Australia, businesses that design or construct machinery for use at work have the same legal obligations in relation to eliminating or minimizing risks and provision of safety information, among other matters. Through a harmonized code of practice, *Managing Risks of Plant in the Workplace*, businesses are encouraged to apply a common set of technical standards, many of which are based on European (CEN) standards adopted by the International Organization for Standardization (ISO) (Safe Work Australia, 2012b; Standards Australia, 2012).

As important, OHS regulators in jurisdictions with harmonized legislation have a consistent range of mechanisms with which to capture manufacturer attention and foster systematic action to address safety in machinery design and construction (Johnstone, Bluff and Clayton, 2012, ch. 8). As well as improvement and prohibition notices, they can issue infringement notices (on-the-spot fines), and accept enforceable undertakings, which are a useful mechanism for addressing firms' commitment, capacity and arrangements for ongoing self-regulation (Johnstone and King 2008). The OHS regulators can also initiate legal proceedings and, if a person is convicted or found guilty of an offence, the courts can impose a fine and/or different types of orders. Such orders are capable of impacting on managerial motivations other than economic ones and can, in effect, require firms to address the factors producing an offence, including weaknesses in their capacity and arrangements to comply with the law (Gunningham and Johnstone, 1999, pp. 256–8; Johnstone, Bluff and Clayton, 2012, pp. 761–4).

It is timely then to contemplate how safety regulators can foster constructive self-regulatory action by machinery manufacturers. This research has pointed to the need for regulators to think strategically about how to engage effectively with firms as designers and manufacturers in order to build capacity, shape motivations and sustain strong performance for substantive safety outcomes.

In Europe there have also been calls to enhance inspection and enforcement of machinery safety, with pressure for more market surveillance and a more harmonized approach to surveillance across member states (Boy and Limou, 2003, pp. 96–8, 107; FEM, 2010; Industry's Support Platform, 2013a; 2013b). Surveillance may be conducted at trade fairs, production facilities, distribution outlets and customs check points or, on an ad hoc basis, after accidents involving machinery (Boy and Limou, 2003, pp. 83–5; HSE, 2013a; 2013b; SWEA, 2002). In principle, such surveillance may result in prohibition of particular machinery, its withdrawal from the market or restrictions on its free trade (European Commission, 2006 recital 10, article 11). However, disparities in member states' legal and administrative frameworks, regulators' resources and technical expertise, regulatory strategy and enforcement mechanisms have given rise to differences in the nature and extent of surveillance, and a high level of non-conformity (Boy and Limou, 2003, pp. 89–98; Industry's Support Platform, 2013b).

The research with Australian manufacturers suggests some strategic directions for compliance support, inspection and enforcement (regulatory interventions). They may also be of interest to European and other regulators seeking to engage more effectively with businesses as manufacturers or suppliers about the safety of the machinery they place on the market.

*Strategic Directions for Compliance Support, Inspection and Enforcement*

It is likely that to give real priority to interventions with upstream parties, regulators will need to establish dedicated policies, strategies and programmes of action that go beyond *routine surveillance* of machinery at events and distribution outlets, and response to incidents or complaints. This will require some rethinking of regulators' analytical frameworks, as immediate injuries and costs are likely to take precedence over risks emerging longer term due to machinery design or construction failures, particularly in the context of risk-based regulation (Baldwin and Black, 2008; Hutter, 2005). A more systematic approach to determining *priorities for regulatory interventions* would be facilitated by establishing national databases (and cross-national if feasible) to record and analyse the types of machinery and components associated with injuries, incidents, complaints and inspectors' directions, notices or other enforcement action.

For interventions with manufacturers, this research points to the merits of a contextual, responsive, networked and principled approach. Improving firm performance will require a change in the mix and balance of constituents of knowledge about machinery safety and motivational factors in firms. As such opportunities lie in a *contextual* element in which regulators analyse the bases from which those involved in machinery design and construction learn about safety matters (such as particular technical standards, customers and suppliers, injury or incident information), and the motivational factors driving or constraining their action to address machinery safety matters. By analysing the contributions to knowledge and the separate strands of motivational factors in firms, regulators can better identify the influences shaping firms responses on machinery safety matters and tailor their interventions accordingly. The aim, in so doing, would be to harness and reinforce influences that support constructive learning and action to address machinery safety matters, and to disarm or at least minimize the impact of those that impede such action.

The *responsive* element involves tailoring regulatory interventions, taking knowledge and motivational factors into account. This approach would use a judicious mix of: cooperation through dialogue, advice, persuasion, warnings and negotiated outcomes; insistence through administrative notices;[7] probation through undertakings which commit firms to implementing particular preventive measures; and deterrence through prosecution (Johnstone, 2004b, pp. 157, 168–74; Johnstone, Bluff and Clayton, 2012, ch. 8).

*Networked* interventions with a cross-section of firms that design, construct, import, supply and purchase machinery within the same markets have the advantage of countering uneven treatment which may engender resistance, especially among firms supplying into competitive markets (see Chapter 5). Such interventions also leverage state efforts by fostering the communication of regulatory messages within supply chains and networks. For the same reasons, and to optimize the use of scarce regulator resources, there is merit in coordinating regulatory interventions across borders, whether that is within Australia, across Europe, or otherwise (Bluff, et al., 2012; and see Bluff and Gunningham, 2012; Boy and Limou, 2003, p. 118; Industry's Support Platform, 2013b). Coordinated and networked interventions also better support a *principled* approach, to the extent that they promote consistency, fair treatment and transparency (see Yeung, 2004, pp. 36–43; and see Braithwaite V, 2009, pp. 82–4; Murphy 2005; Tyler 1997).

As many study firms did not manage machinery safety matters systematically (see Chapter 7), a regulatory emphasis on securing firms' commitment, capacity and arrangements to self-regulate on an ongoing basis is indicated. Such arrangements need not be complex management systems but should sustain continuing action to address safety in machinery design and construction, rather than ad hoc reactions to incidents, customer requirements, complaints or other stimuli. Especially relevant are processes and resources for recognizing hazards and determining risk control measures, obtaining input from customers and users at the design stage, and receiving and acting on after-market reports of incidents and hazards arising in end use. Also crucial is building capacity to address safety matters in machinery design and construction, which is facilitated by firms engaging persons with health and safety know-how so as to provide a conduit to regulatory and professional communities of practice and bodies of knowledge, and to help frame issues for attention and improved risk management (see Chapter 6, and see Broberg and Hermund, 2007; Jensen, 2002b; Nytrö, Saksvik and Torvatn, 1998; Parker, 2002, ch. 5).

---

7   For example, infringement, improvement and prohibition notices.

A focus on processes and resources to sustain systematic attention to safety does not, however, imply that regulators would confine themselves to inspecting whether firms have particular arrangements in place. This research has provided ample evidence that procedural action such as conducting risk assessments does not necessarily ensure compliance with the regulatory goal of prevention (see Chapter 7). Regulators will also need to examine firms' performance against substantive outcomes so that people using or encountering machinery will be protected from risks and informed about them.

What types of programmes would lend themselves to these sort of contextual, responsive, networked and principled designs, and a focus on fostering commitment, capacity and arrangements to comply, as well as checking firms' performance for substantive safety outcomes? In broad terms, programmes that involve regulators, across jurisdictions, communicating, inspecting and enforcing with firms that design, manufacture, import and supply *particular types of machinery* or supply *machinery for particular industry sectors* (agriculture, construction, food processing, for example). Such targeted programmes might include provision of guidance material and self-audit tools to assist firms to self-regulate, and follow-up visits to firms (or a random sample of firms), to assess their action and require remedial action if needed. (For programmes with elements of this approach see Boy and Limou (2003, p. 98), HSE (2013a; 2013b), HWSA (2007), SWEA (2002) and see Industry's Support Platform (2013b)).

Other programmes could target machinery identified as *high risk* and warranting a problem solving approach (Sparrow, 2000), to address problems which have become intractable due to competing functionality, competition or other commercial goals. This type of programme is likely to begin with regulators clarifying, through research and consulting with end users and industry stakeholders, the nature of the risks, possible design and redesign solutions and other measures to minimize risks. Typical programme elements would be the production and dissemination of guidance material, and inspection and enforcement with relevant firms that design, construct, import or supply the machinery, as well as key customers. There may also be a need to negotiate amendments to relevant technical standards. Australian regulatory interventions for forklifts and quad bikes are examples of programmes of this type (Lambert and Associates, 2003; Safe Work Australia, 2013c; Skinner and Stewart, 2006; WorkSafe Victoria, 2003a; 2003b).

A critical element in any intervention is for regulators to ensure that firms and representative stakeholders (industry associations and unions) are clear how and why regulators take different types of enforcement action. Where regulators use higher order mechanisms, such as enforceable undertakings or prosecution, they would communicate the implications of such actions to relevant firms and stakeholders, and check that industry media and networks accurately relay the implications to their constituents (on the spread of misinformation see Chapter 5).

The examples above suggest various ways that regulators might engage with upstream parties for machinery more extensively and in a more meaningful way, but can regulators do more than this? One way they could broaden their influence would be to encourage professional, educational and industry stakeholders to collaborate in providing experiential learning programmes, recognizing the extent to which key decision makers in design and construction learn about safety matters through practice. These ideas about capacity building are explored in the next section.

*Strategic Directions for Capacity Building*

The study firms represented a community of practice (Lave and Wenger, 1990; Palincsar, 1998) that was largely separate from regulatory and specialist (safe design) communities of practice. If manufacturers remain disconnected from the regulatory community of practice[8] they lack a legally grounded conceptual framework for managing safety in machinery design and construction. If they fail to engage with the specialist community of practice, invaluable information about designing and constructing inherently safe machinery, and effectively informing end users is not part of their frame of reference.

A way forward is for regulators to join with professional and industry associations, and educators (for example, for engineering, safety and metal trades) to foster programmes to help build capacity for the safe design and construction of machinery among individuals with different professional and vocational backgrounds. The initiatives proposed here would extend and adopt a fundamentally different approach from past offerings; the guidance materials and training manuals produced for use in engineering or other educational programmes (see for example ASCC, 2006a; 2006b; NIOSH, 2013). They would provide practical opportunities for participants to learn about addressing safety in the course of design and construction activities, and what it means to comply with their legal obligations. That is, capacity-building initiatives would be practice-based. They would be structured and organized around authentic work activities, supported and guided by experienced practitioners, and would provide opportunities to reinforce and hone knowledge and skills (Billett, 2010, chs. 1, 4; n.d.). To widen the impact, such initiatives could also be made available to health and safety advisers and consultants, producers of safety components, inspectors and other influential actors.

With regard to key elements of programmes, participants would learn how to ensure that risks have been eliminated or minimized (so far as reasonably practicable), how to effectively test and examine machinery, and how to produce informative and user-friendly safety information. Special attention would be paid to challenging the unsafe worker attitude, which this research has shown plays a defining role in justifying firms' decisions to limit or not implement preventive measures.

The research also suggests the merits of participants learning about practices linked with better performance for substantive safety outcomes (see Chapters 6

---

8   On communities of practice see Lave and Wenger (1990) and Palincsar (1998).

and 7). This would mean learning about how to interact effectively with end users at the design stage, trial prototypes or models of machinery with end users, and inspect the end use work environment in order to recognize risks for the people who will work with or in the vicinity of the machinery. Other core practices are ways for manufacturers to ensure they receive information about injuries, incidents and hazards, and make good use of this information in redesigning machinery, taking remedial action, and reviewing safety information and testing regimes. In addition, the research suggests that manufacturers could make better use of component suppliers for state-of-the-art control measures, not just industry standard ones, and better recognize the strengths of and limits to technical standards applicable to their machinery.

Capacity building initiatives can support better execution of these feasible practices by making links to specialist information resources and assisting participants to make effective use of human factors or safety engineering professionals who have knowledge to support safe design and construction. Finally, capacity building initiatives will need to compete with the ongoing, day-to-day constituents of participants' knowledge about machinery safety. To this end, it may be beneficial to design initiatives to include follow up learning opportunities and establish support networks, in order to foster continuing exchange of information and reinforce efforts to manage safety in machinery design and construction.

## Conclusion

This book contributes to a growing body of literature about business performance for safety and compliance with state imposed legal obligations. Based on empirical research with machinery manufacturers and OHS regulators, the book advances understanding about how multiple state and non-state influences shape business conduct. From a starting point which demonstrated the mixed and often mediocre performance of manufacturers for hazard recognition, risk control and provision of safety information, the book has revealed the highly contextualized and complex nature of motivational factors and knowledge about machinery safety matters in manufacturing firms, and hence their rather idiosyncratic action and performance for substantive safety outcomes.

The book has also illuminated several of the conceptual themes central to a holistic and plural explanation of business conduct and compliance (Parker and Nielsen, 2011). It has elucidated the processes by which regulatees construct knowledge as a key element of capacity, the interplay between co-existing and sometimes conflicting motivational factors in firms, and the contribution of state and non-state actors to each of these. Crucially, the law and state regulators take their place among the diverse bases from which boundedly rational decision makers construct knowledge, and their competing motivations, values and attitudes.

The empirical findings and theorizing from this research have important implications for policy and professional interventions requiring or encouraging the safe design and construction of machinery, or the achievement of other social and regulatory goals. In essence, if such interventions are to claim a place in the complex mix of influences shaping business conduct, regulators and policy makers, specialists, industry and educational stakeholders will need to implement carefully crafted and focused interventions, which are designed to capture business attention, build capacity, minimize resistance but also take on persistent poor performance. The book has fashioned some strategic directions which are designed to achieve this.

# Appendix
# Research Design and Methodology

The epistemology underlying this research was constructionism: the view that all meaningful reality is contingent upon human practices being constructed in and out of interactions between people and their world, and developed and transmitted within a social context (Crotty, 1998, pp. 8–9, 42–5; Flick, 2006, pp. 78–9). A qualitative research design and methodology was applied, as the research sought to learn from manufacturers and regulators about their experiences, perceptions and practices, and explore in depth how and why things happen (Mason, 1996, p. 4; Marshall and Rossman 2006, pp. 53–4; Miles and Huberman, 1994, pp. 7, 10; Morse and Richards, 2002, pp. 27–8).

The aim of the research was to examine how firms that designed and constructed machinery for use at work addressed safety matters for the machinery they produced, and the factors shaping their responses. The research explored these issues through empirical studies with manufacturing firms and with Australian occupational health and safety (OHS) regulators, and a legal review and analysis of relevant legal obligations and case law. These different perspectives enabled triangulation through the use of different sources, types of data and methods to corroborate, challenge and illuminate insights on particular topics of interest (Berg, 2007, pp. 5–8; Denzin and Lincoln, 1998, p. 4; Flick, 2007, pp. 389–90).

The author conducted all parts of this research. From a background in biological science Elizabeth Bluff worked as a health and safety professional for 20 years in OHS management, standard setting and policy roles, before making the transition to academia and conducting research in the inter-related fields of regulation and compliance, safe design, and risk and safety management. Before commencing empirical work with machinery manufacturers she completed special purpose training in hazard recognition and risk assessment for machinery. Her formal qualifications are a Bachelor of Science (Hons), Master of Applied Science (OHS) and Doctor of Philosophy (PhD) in OHS regulation.

## The Legal Review and Analysis

The legal review and analysis clarified the principal legal obligations for the safe design and construction of machinery, which applied to firms producing machinery in Australia and supplying into international as well as local markets. In focus were the prevention and enforcement provisions in Australian OHS law; prosecutions of machinery designers, manufacturers and suppliers by Australian OHS regulators; and the technical standards relevant to machinery safety and risk management, which had mandatory or evidentiary status under OHS law. Also examined was the European regulatory regime for machinery safety implementing the *Machinery Directive* and relevant harmonized standards (European Commission, 1998a; 1998b; 1998c); this being the most influential regulatory regime for Australian machinery manufacturers and in global markets more generally (European Commission, 2014; European Standards Organization, 2009; IMS Research, 2009). The potential for legal action under the common law was considered with regard to civil actions for the torts of negligence and breach of statutory duty concerning machinery design and construction. The legal review informed the development of data collection instruments for the studies with machinery manufacturers and with OHS regulators, and the frames for analysing the performance of manufacturers and the activities of OHS regulators.

## The Empirical Studies

*Methods and Instruments*

In-depth, face-to-face interviews were conducted at study firms' premises and the offices of OHS regulators. Accepted principles for qualitative interviews were applied (Berg, 2007, pp. 99–105, 130–1; Gillham, 2000, pp. 28–36, 40–50; Minichiello, et al., 1995, pp. 73–4). Semi-structured schedules were used to ask about consistent topics through open-ended questions, in order to elicit detailed responses from participants. Interviews were audio taped and transcribed verbatim because the ways participants articulated their knowledge, experience, understandings and motivations were also of interest (Mason, 1996, p. 55).

In study firms, interviews were conducted with key individuals who were the main decision makers in their firms for machinery design and construction, and had knowledge and experience of how safety was addressed in the course of these activities. These interviewees were asked general questions about their experience and qualifications, the firm's operations (the machinery produced, how it was designed and made), and the markets for the firm's machinery (the industries and locations supplied in Australia and other countries). Interviewees were then asked how machinery safety was addressed including the firm's actions, practices and processes for design and construction, and factors and mechanisms motivating or constraining attention to safety matters. They were also asked about sources of knowledge about safety, awareness and understanding of relevant legal obligations, the firm's experience of inspection and enforcement, risk management, testing and examination, provision of safety information, and any arrangements for managing safety in machinery design and construction.

Regulator interviewees were asked general questions about their qualifications and experience before more specific questions about their knowledge of the legal obligations of machinery designers and manufacturers, and their role in providing advice and information, inspection and investigation, issuing notices and prosecution, determining compliance, design notification and registration. They were also asked for their assessment of the extent of compliance, and their perceptions of any problems with OHS law, inspection and enforcement and suggestions for improvement.

Access to various forms of documentation and audio–visual materials was requested at the time of both the manufacturer and the regulator interviews. Whenever possible, copies were obtained for subsequent review. Otherwise documents were reviewed and notes recorded *in situ*. For manufacturers, examples of documents and materials included product brochures, CDs, videos, websites, machinery safety information, technical standards and other information resources, risk assessments, design specifications, design verification documentation, and records of safety testing or examination. For regulators, examples included information and guidance materials, policies and procedures, strategy papers, notices issued, prosecution files and records. These documents might be available as print materials or through the regulators' websites. Documentation and materials depicted reality for manufacturers and regulators, as well as providing evidence of the depth and quality of activity and experience within an organization (Silverman, 2001, pp. 123, 128).

In the manufacturer study each firm's machinery was observed. Specifically this involved identifying potential sources of harm, whether risk control measures were incorporated, and the type and adequacy of these, and any decals or signage. When possible, firms' design and construction facilities and practices were also observed, including project design teams at work and the software or methods used by them, and testing and examination of machinery. Notes were recorded immediately after leaving the manufacturer's premises (Richards, 2005, p. 39).

All empirical materials (interview transcripts, documentation and records of observation) were de-identified so that study firms, individual participants in these firms, individual staff of OHS regulators, and other organizations or individuals named by these participants were known only to the researchers. In addition, consent forms and details of coding systems used in de-identifying these empirical materials were kept secure and separate from de-identified materials.

*Sampling and Participation for the Manufacturer Study*

The study population was manufacturing firms in two Australian states, Victoria and South Australia. Three criteria were applied in defining potential participants in this population. The firms must be involved in the design and construction of machinery, even if they outsourced some aspects of design, component production or supply, or assembly. They must produce machinery for use at work, either exclusively or as part of their market, and the machinery must be power driven. Taking these three criteria into account, firms producing a wide range of machinery were potential candidates for inclusion in the study population.

The Victorian and South Australian workers' compensation agencies provided lists of businesses classified as manufacturers of industrial machinery (WorkCover Corporation, 2000; Victorian WorkCover Authority, 2003c), together with the business location and remuneration of each firm as an indicator of size. These lists, after checking for duplication and erroneous inclusions,[1] constituted the sampling frames (Mason, 1996, p. 102; Neuman, 1997, p. 203) for selecting firms in the two study states.

The stratified, purposive sampling strategy used to select manufacturers was theoretically driven by a concern to capture major variations and to see different instances of the phenomena studied (Mason, 1996, pp. 92–3; Miles and Huberman 1994, pp. 27–9; Morse and Richards, 2002, p. 173). The strategy involved identifying strata within each sampling frame according to some key characteristics and attributes: state of operation (Victoria or South Australia); location of the business within the state (metropolitan or regional); and firm size – small (<20 employees), medium (20–99 employees), and large (100 or more employees). Other factors of interest, such as the type and complexity of the machinery, whether it was custom made or produced as standard models, and firms' markets could not be determined reliably prior to data collection. They were explored in interviews.

Firms were randomly selected from each stratum and contacted to arrange participation. Proportionally more small and medium firms were selected reflecting the characteristics of firms in the study population (ABS, 2001). Sampling was then to the point of saturation, when no new information was forthcoming from interviewees, and the data were of sufficient depth and scope (Flick, 2006, pp. 127–8; Richards, 2005, pp. 19–20, 135–6). Table A1.1 below presents the sample of firms participating in the research, by state, firm size and location within the state. In Victoria, of the participation rate was 70 per cent (32/46), and in South Australia the participation rate was 69 per cent (34/49).

**Table A1.1    Distribution of manufacturing firms across the sampling strata**

|  | <20 | | 20–99 | | 100+ | | |
|---|---|---|---|---|---|---|---|
|  | Met | Reg | Met | Reg | Met | Reg | Total |
| South Australia | 12 | 6 | 8 | 4 | 4 | 0 | 34 |
| Victoria | 12 | 4 | 8 | 4 | 4 | 0 | 32 |
| Total | 24 | 10 | 16 | 8 | 8 | 0 | 66 |

*Note:* in Table A1.1, Met is a metropolitan (capital city) location, Reg is a regional (country) location and '0' indicates no firms in sampling frame for this stratum.

1    For example the firm maintained or serviced machinery, or traded in component parts rather than being an end product manufacturer.

In each firm, a senior manager authorized participation and provided informed consent to the researcher conducting interviews, document review and observation. Interviewees were usually directors, owners or senior managers with responsibility for engineering or production, but some were persons responsible for engineering, technical, health and safety, or other specialist functions. Prior to each interview the individual participant was provided with a written outline of the research, including the arrangements to protect confidentiality and secure data, and their signed, informed consent to participate was obtained.

*Sampling and Participation for the Regulator Study*

For document analysis, information and materials were collected from the websites of all Australian OHS regulators. For interviews, the study focused on the regulators in the two states where the study firms based their operations (Victoria and South Australia). Senior officers of the respective agencies authorized their organizations' participation, providing informed consent to the researcher conducting interviews with a range of staff and having access to relevant documentation. The sampling frame for each regulator was comprised of the staff in a job role in which they would encounter the regulator's inspection and enforcement policy, strategies and practices for machinery design and construction.

A stratified, purposive sampling strategy was used to select interview participants. This involved identifying strata according to some key attributes reflecting different types of job roles and experiences as directors or executive level officers; managers of strategies, programmes or field operations; technical specialists (engineers and ergonomists); legal investigators and field inspectors. The sample of 32 regulator participants is presented in Table A1.2.

**Table A1.2 Distribution of regulator interviewees across the sampling strata**

|  | **Director** | **Manager** | **Specialist** | **Legal** | **Inspector** | **Total** |
|---|---|---|---|---|---|---|
| South Australia | 2 | 2 | 3 | 0 | 9 | 16 |
| Victoria | 2 | 4 | 3 | 1 | 6 | 16 |
| Total | 4 | 6 | 6 | 1 | 15 | 32 |

*Note:* in Table A1.2, '0' indicates no staff in sampling frame for this stratum.

Interviewees were randomly selected from each stratum and contacted to arrange participation. Sampling was to the point of saturation, when no new information was forthcoming. The overall participation rate was very high (Victoria: 94 per cent (16/17); South Australia: 89 per cent (16/18)).

Prior to interview each participant was provided with a written outline of the research, including the arrangements to protect confidentiality and secure data, and their signed, informed consent to participate was obtained. Approximately half were current inspectors with operational experience of interactions with businesses (15/32), but taking into account past experience, two thirds had such operational experience (20/32).

### Data Analysis, Explanation and Theory Building

Qualitative methods were applied to analyse empirical data and to construct explanation and theory out of (grounded in) the data, by applying inductive reasoning and using the data to corroborate, challenge, illuminate or illustrate insights (Denzin and Lincoln 2003). In part this explanation was descriptive, providing an account of what was going on. It also involved comparison of phenomena, processes, meanings and other aspects of the data (Mason, 1996, p. 137; Richards, 2005, pp. 129–31). The theory constructed was substantive or local theory, to explain the phenomena studied and make sense of the data (Neuman, 1997, pp. 47–8, 55–6; Richards, 2005, pp. 129–31; Silverman, 2001, pp. 3–4).

Core analytic procedures applied were immersion in and systematic reflection on the data, recording emerging concepts and themes, and generating categories and coding data to categories to enable retrieval and analysis of all data on particular topics. These slices of data were used to develop descriptions of particular practices or phenomena, and to examine similarities, differences and patterns across data for a particular topic. Other core processes were exploring co-occurrences and reflecting upon links between topics, conceptual and thematic categories, and characteristics and attributes;[2] and interpreting and developing explanation to account for those similarities, differences, patterns and links across the data (Gillham, 2000, pp. 63–6; Marshall and Rossman, 2006, pp. 156–8; Neuman 1997, pp. 426–8; Richards 2005, pp. 69–70, 96).

The purpose of looking for co-occurrences was not to identify causal patterns or relationships in a statistical sense but to support a process of analytic induction and development of plausible explanations (Mason 1996, pp. 88, 118, 143–4). Substantive theory was constructed in the form of an integrative interpretation that made sense of the data, and brought meaning and coherence to themes, patterns and relationships in the data (Marshall and Rossman, 2006, pp. 161–2; Morse and Richards 2002, pp. 169–70; Neuman, 1997, pp. 46–8; Richards, 2005, pp. 128–34; Silverman, 2001, pp. 237–40). The method for constructing theory involved a recursive cycle of inductive reasoning from the data, and deductive reasoning by applying insights from the literature.

---

2   Firm characteristics and attributes included size and location, type of machinery produced and markets. Individual attributes included key individuals' and designers' qualifications and experience.

# References

ABS see Australian Bureau of Statistics.

Adams, P., 1999. Application in general industry. In: W. Christensen and F. Manuele, eds. 1999. *Safety through design*. USA: National Safety Council Press. pp. 155–69.

Alexander, P., 1991. Coming to terms: how researchers in learning and literacy talk about knowledge. *Review of Educational Research*, 61(3), pp. 315–43.

Al–Tuwaijri, S., Fedotov, I., Feitshans, I., Gifford, M., Gold, D., Machida, S., Nahmias, M., Niu, S. and Sand, G., 2008. *Beyond death and injuries: The ILO's role in promoting safe and healthy jobs*. Geneva: International Labour Organization.

American National Standards Institute, 2005. *ANSI/AIHA Z10: 2005 Occupational health and safety management systems*. Washington: American National Standards Institute.

American Society of Safety Engineers, 2011. Technical brief on ANSI/ASSE Z590.3–2011, Prevention through design. Guidelines for addressing occupational hazards and risks in design and redesign processes. [online] American Society of Safety Engineers. Available at: <https://www.assp.org/standards> [Last accessed 27 August 2014].

ANSI see American National Standards Institute.

ASCC see Australian Safety and Compensation Council.

ASSE see American Society of Safety Engineers.

Association of Workers' Compensation Boards of Canada, 2012. *National work injury and disease fatality statistics 2010–2012*. Ontario: Association of Workers' Compensation Boards of Canada.

Atherley, G., 1975. Strategies in health and safety at work. *The Production Engineer*, 54, pp. 50–5.

Atherley, G., 1978. *Occupational health and safety concepts*. London: Applied Science.

Australian Bureau of Statistics, 2001. *Management unit counts by ANZSIC and employment range*. Canberra: Australian Bureau of Statistics.

Australian Government, 2008. *National review into model occupational health and safety laws. First report to the Workplace Relations Ministers' Council*. Canberra: Commonwealth of Australia.

Australian Government, 2009. *National review into model occupational health and safety laws. Second report to the Workplace Relations Ministers' Council*. Canberra: Commonwealth of Australia.

Australian Government, n.d. *Remedies for breach of contract.* [online] Australian Government. Available at: <http://www.ag.gov.au/Consultations/Documents/ReviewofAustraliancontractlaw/Remediesforbreachofcontract.pdf> [Last accessed 27 August 2014].

Australian Safety and Compensation Council, 2006a. *Guidance on the principles of safe design for work.* Canberra: Commonwealth of Australia.

Australian Safety and Compensation Council, 2006b. *Safe design for engineering students. An educational resource for undergraduate engineering students.* Canberra: Commonwealth of Australia.

Australian Safety and Compensation Council, 2008. *Guidelines for integrating OHS into national industry training packages.* Canberra: Commonwealth of Australia.

AWCBC see Association of Workers' Compensation Boards of Canada.

Ayres, I. and Braithwaite, J., 1992. *Responsive regulation. Transcending the deregulation debate.* Oxford: Oxford University Press.

Backstrom, T. and Döös, M., 1997. The technical genesis of machine failures leading to occupational accidents. *International Journal of Industrial Ergonomics*, 19(5), pp. 361–76.

Backstrom, T. and Döös, M., 2000. Problems with machine safeguards in automated installations. *International Journal of Industrial Ergonomics*, 25(6), pp. 573–85.

Badham, R. and Ehn, P., 2000. Tinkering with technology: human factors, work redesign, and professionals in workplace innovation. *Human Factors and Ergonomics in Manufacturing*, 10(1), pp. 61–82.

Baggs, J., Silverstein, B. and Foley, M., 2003. Workplace health and safety regulations: impact of enforcement and consultation on workers' compensation claims rates in Washington state. *American Journal of Industrial Medicine*, 43(5), pp. 483–94.

Baker, W., Temperley, J., Hawkins, A. and Fragar, L., 2005. *Farm machinery guarding guidance material.* Moree: Australian Centre for Agricultural Health and Safety.

Baldwin, R., 1995. *Rules and government.* Oxford: Oxford University Press.

Baldwin, R. and Black, J., 2008. Really responsive regulation. *Modern Law Review,* 71(1), pp. 59–94.

Baldwin, R. and Cave, M., 1999. *Understanding regulation. Theory, strategy and practice.* Oxford: Oxford University Press.

Baldwin, R., Scott, C. and Hood, C., 1998. *A reader on regulation.* Oxford: Oxford University Press.

Bamberg, S. and Boy, S., 2008. *The new Machinery Directive. A tool to uncover the changes introduced by the revised directive.* [online] European Trade Union Institute. Available at: <http://www.etui.org/Publications2/Guides> [Last accessed 27 August 2014].

Baram, M., 2006. *Liability and its influence on designing for product and process safety.* Boston : Boston University School of Law.

Bardach, E. and Kagan, R., 2002. *Going by the book. The problem of regulatory unreasonableness*. New Jersey: Transaction Publishers.

Benedyk, R. and Minister, S., 1998. Applying the BeSafe method to product safety evaluation. *Applied Ergonomics,* 29(1), pp. 5–13.

Berg, B., 2007. *Qualitative research methods for the social sciences*. 6th ed. Boston: Pearson.

Billett, S., 1996. Situated learning: bridging sociocultural and cognitive theorising. *Learning and Instruction* 6(3), pp. 263–80.

Billett, S, 2001. Knowing in practice: re–conceptualising vocational expertise. *Learning and Instruction,* 11(6), pp. 431–52.

Billett, S., 2003. Sociogeneses, activity and ontogeny. *Culture Psychology,* 9(2), pp. 133–69.

Billett, S., 2006. Relational interdependence between social and individual agency in work and working life. *Mind, Culture and Activity,* 13(1), pp. 53–69.

Billett, S., 2008a. Learning throughout working life: a relational interdependence between personal and social agency. *British Journal of Educational Studies,* 56(1), pp. 39–58.

Billett, S., 2008b. Emerging perspectives on workplace learning. In: S. Billett, C. Harteis and A. Eteläpelto eds. 2008. *Emerging perspectives of workplace learning*. Rotterdam: Sense Publishers. pp: 1–15.

Billett, S. ed., 2010. *Learning through practice. Models, traditions, orientations and approaches*. Dordrecht: Springer.

Billett, S., n.d. *Realising the educational worth of integrating work experiences in higher education*. [online] Griffith University. Available at: <http://www98.griffith.edu.au/dspace/bitstream/handle/10072/29123/58192_1.pdf?sequence=1> [Last accessed 27 August 2014].

Black, J., 1997. *Rules and regulations*. Oxford: Clarendon Press.

Black, J., 2001a. Decentring regulation: understanding the role of regulation and self–regulation in a 'post–regulatory' world. In: M. Freeman ed. 2001. *Current legal problems 2001, volume 54*. Oxford: Oxford University Press. pp. 101–44.

Black, J., 2001b. Managing discretion. In: Australian Law Reform Commission, *Conference on penalties, principles and practice in government regulation*. Sydney, Australia, June 2001. Sydney: Australian Law Reform Commission.

Black, J. and Baldwin, R., 2010. Really responsive risk–based regulation. *Law and Policy*, 32(2), pp. 181–213.

Blewett, V., 2001. *Working together: a review of the effectiveness of the health and safety representative and workplace health and safety committee system in South Australia*. Adelaide: WorkCover Corporation of South Australia.

Bluff, L., 2001. Producing risks: creating safety – how is product safety addressed in management systems? In: W. Pearse, C. Gallagher and L. Bluff eds. 2001. *Occupational health and safety management systems*. Melbourne: Crown Content. pp. 101–21.

Bluff, L., 2004. Regulating the safe design of plant. *The Journal of Occupational Health and Safety, Australia and New Zealand*, 20(3), pp. 229–39.

Bluff, L. and Gunningham, N., 2004. Principle, process, performance or what? New approaches to OHS standards setting. In L. Bluff, N. Gunningham and R. Johnstone eds. 2004. *OHS regulation for a changing world of work*. Sydney: The Federation Press. pp. 12–42.

Bluff, E. and Gunningham, N., 2012. Harmonising work health and safety regulatory regimes. *Australian Journal of Labour Law*, 25(2), pp. 85–106.

Bluff, E. and Johnstone, R., 2005. The relationship between 'reasonably practicable' and risk management regulation. *Australian Journal of Labour Law*, 18(3), pp. 197–239.

Bluff, E., Johnstone, R., McNamara, M. and Quinlan, M., 2012. Enforcing upstream: Australian health and safety inspectors and upstream duty holders. *Australian Journal of Labour Law*, 25(1), pp. 23–42.

Bohle, P. and Quinlan, M. eds, 2000. *Managing occupational health and safety. A multidisciplinary approach*. 2nd ed. Melbourne: MacMillan Publishers Australia.

Boston, O., Culley, S. and McMahon, C., 1999. Life-cycle management of supplier literature: the pertinent issues. *Journal of Product Innovation Management*, 16(3), pp. 268–81.

Boy, S. and Limou, S., 2003. *The implementation of the machinery directive. A delicate balance between market and safety*. Brussels: European Trade Union Technical Bureau for Health and Safety.

Braithwaite, J., 2011. The essence of responsive regulation. *University of British Columbia Law Review*, 44(3), pp. 475–520.

Braithwaite, V., 1995. Games of engagement: postures within the regulatory community. *Law and Policy*, 17(3), pp. 225–55.

Braithwaite, V., 2009. *Defiance in taxation and governance. Resisting and dismissing authority in a democracy*. Cheltenham: Edward Elgar.

Braithwaite, V., Braithwaite, J., Gibson, D. and Makkai, T., 1994. Regulatory styles, motivational postures and nursing home compliance. *Law and Policy*, 16(4), pp. 363–94.

Braithwaite, V., Murphy, K. and Reinhart, M., 2007. Taxation threat, motivational postures, and responsive regulation. *Law and Policy*, 29(1), pp.137–58.

Brauer, R., 1994. *Safety and health for engineers*. 1st ed. New Jersey: John Wiley and Sons.

Brauer, R., 2006. *Safety and health for engineers*. 2nd ed. New Jersey: John Wiley and Sons.

Brewer, J. and Hsiang, S., 2002. The ergonomics paradigm: foundations, challenges and future directions. *Theoretical Issues in Ergonomics Science*, 3(3), pp. 285–305.

British Standards Institution, 2007. *BS OHSAS 18001: 2007 Occupational health and safety management systems – requirements*. London: British Standards Institution.

British Standards Institution, 2008. *BS OHSAS 18001: 2008 Occupational health and safety management systems – guidelines for the implementation of OHSAS 18001*. London: British Standards Institution.

Broberg, O., 1997. Integrating ergonomics into the product development process. *International Journal of Industrial Ergonomics*, 19(4), pp. 317–27.

Broberg, O., 2007. Integrating ergonomics into engineering: empirical evidence and implications for ergonomists. *Human Factors and Ergonomics in Manufacturing*, 17(4), pp. 353–66.

Broberg, O. and Hermund, I., 2007. The OHS consultant as a facilitator of learning in workplace design processes: Four explorative case studies in current practice. *International Journal of Industrial Ergonomics*, 37(9–10), pp. 810–16.

Brooks, A., 1993. *Occupational health and safety law in Australia*. 4th ed. Sydney: CCH Australia.

Brown, J. S., Collins, A. and Duguid, P., 1989. Situated cognition and the culture of learning. *Educational Researcher*, 18(1), pp. 32–42.

Brown, J.S. and Duguid, P., 1991. Organizational learning and communities-of-practice: toward a unified view of working, learning and innovation. *Organization Science*, 2(1), pp. 40–57.

Bruseberg, A. and McDonagh–Philp, D., 2002. Focus groups to support the industrial/product designer: a review based on current literature and designers' feedback. *Applied* Ergonomics, 33(1), pp. 27–38.

BSI see British Standards Institution.

Busby, J., 2003. *HSE RR 054: 2003 Mutual misconceptions between designers and operators of hazardous installations*. Norwich: HMSO.

Butters, L. and Dixon, R., 1998. Ergonomics in consumer product evaluation: an evolving process. *Applied Ergonomics*, 29(1), pp. 55–8.

Canadian Standards Association, 2006. *CS Z1000: 2006 Occupational health and safety management*. Toronto: Canadian Standards Association.

Caple, D., 2000. *Assessment of policy implications arising from research undertaken for the safe design project*. Sydney: National Occupational Health and Safety Commission.

CEN see European Committee for Standardization.

CENELEC see European Committee for Electrotechnical Standardization.

Clarke, P., Clarke, J. and Courmadias, N., 2005. *Butterworths casebook companions. Contract law*. Sydney: LexisNexis Butterworths.

Cordero, C. and Muñoz Sanz, J., 2009. Measurement of machinery safety level: European framework for product control. Particular case: Spanish framework for market surveillance. *Safety Science*, 47(1), pp. 1285–96.

Cordero, C., Muñoz Sanz, J., Otero, J. and Muñoz Guijosa, J., 2013. Measurement of machinery safety level in the European market: characterisation of the compliance within the scope of MD98/37/EC. *Safety Science*, 51(1), pp. 273–83.

Corlett, E. and Clark, T., 1995. *The ergonomics of workspaces and machines. A design manual*. 2nd ed. London: Taylor and Francis.

Cowley, S., 2006. *OH&S in small business: influencing the decision makers: the application of a social marketing model to increase the uptake of OHS risk control.* Ballarat: University of Ballarat.

Cowley, S., Culvenor, J. and Knowles, J., 2000. *Safe design project. Review of literature and review of initiatives of ohs authorities and other key players.* Sydney: National Occupational Health and Safety Commission.

Crabb, R., 2000. *HSE RR 306: 2000 Health and safety in the agricultural engineering design process.* Norwich: HMSO.

Creighton, B. and Rozen, P., 2007. *Occupational health and safety in Victoria.* 2nd ed. Sydney: The Federation Press.

Cross, J., 2001. Limitations and use of job safety analysis. *Safety in Australia,* 23(6), pp. 11–12.

Cross, J., Bunker, E., Grantham, D., Connell, D. and Winder, C., 2000. Identifying, monitoring and assessing occupational hazards. In: P. Bohle and M. Quinlan eds. 2000. *Managing occupational health and safety. A multidisciplinary approach,* Melbourne: MacMillan Publishers Australia. pp: 364–427.

Crotty, M., 1998. *The foundations of social research, meaning and perspective in the research process.* Sydney: Allen and Unwin.

CSA see Canadian Standards Association.

Davies, M. and Malkin, I., 2003. *Torts.* 4th ed. Sydney: LexisNexis Butterworths.

Dejoy, D., 1994. Managing safety in the workplace: an attribution theory analysis and model. *Journal of Safety Research,* 25(1), pp. 3–17.

Denzin, N. and Lincoln, Y. eds, 1998. *Strategies of qualitative inquiry.* Thousand Oaks: Sage Publications.

Denzin, N. and Lincoln, Y. eds, 2003. *Collecting and interpreting qualitative materials.* 2nd ed. Thousand Oaks: Sage Publications.

Department of Industry, 2014. *European Community–Australia mutual recognition agreement.* [online] Australian Government. Available at: <http://www.industry.gov.au/industry/TradePolicies/TechnicalBarrierstoTrade/Pages/ECAustMRA.aspx> [Last accessed 27 August 2014].

Department of Trade and Industry, 1999. *Guidelines on the appointment of UK notified bodies to undertake inspection and certification for the purposes of the conformity assessment procedures in the UK regulations.* London: DTI.

Diver, C., 1983. The optimal precision of administrative rules. *Yale Law Journal,* 65(1), pp. 65–109.

Dodge, D., 2001. Basic elements of a product safety program. *Professional Safety,* 46(10), pp. 43–7.

Driscoll, T., Harrison, J., Bradley, C. and Newson, R., 2005. *Design issues in work-related serious injuries.* Canberra: Commonwealth of Australia.

Driscoll, T., Harrison, J., Bradley, C. and Newson, R., 2008. The role of design issues in work-related injuries in Australia 1997–2002. *Journal of Safety Research,* 39(2), pp. 209–14.

DTI see Department of Trade and Industry.

Eraut, M., 2010. Knowledge, working practices and learning. In: S. Billett ed. 2010. *Learning through practice. Models, traditions, orientations and approaches*, Dordrecht: Springer. pp. 37–58.

European Commission, 1989. Council Directive 89/392/EEC of 14 June 1989 on the approximation of laws of the Member States relating to machinery. *Official Journal of the European Union*, L 184, 29/06/1989.

European Commission, 1992. Council Directive 92/58/EEC of 24 June 1992 on the minimum requirements for the provision of safety signs and/or health signs at work (ninth individual Directive within the meaning of Article 16(1) of Directive 89/391/EEC). *Official Journal of the European Union*, L 245, 26/8/1992.

European Commission, 1995. Council Directive 95/16/EC of 29 June 1995 on the approximation of the laws of the Member States relating to lifts. *Official Journal of the European Union*, L 213, 7/9/1995.

European Commission, 1997. Council directive 97/23/EC of 29 May 1997 on the approximation of the laws of the member states concerning pressure equipment. *Official Journal of the European Union*, L 181, 09/07/1997.

European Commission, 1998a. Council directive 98/37/EC of 22 June 1998 on the approximation of laws of the member states relating to machinery. *Official Journal of the European Union*, L 207, 23/07/1998.

European Commission, 1998b. *Community legislation on machinery. Comments on Council Directive 98/37/EC*. Belgium: Office for Official Publications of the European Communities.

European Commission, 1998c. *Useful facts in relation to the machinery directive*. Belgium: Office for Official Publications of the European Communities.

European Commission, 2000a. *Guide to the implementation of directives based on the new approach and the global approach*. Luxembourg: Office for Official Publications of the European Communities.

European Commission, 2000b. Council Directive 2000/14/EC of the European Parliament and of the Council of 8 May 2000 on the approximation of the laws of the member states relating to the noise emission in the environment by equipment for use outdoors. *Official Journal of the European Union*, L 162, 3/7/2000.

European Commission, 2006. Council Directive 2006/42/EC of the European Parliament and of the Council of 17 May 2006 on machinery. *Official Journal of the European Union*, L 157, 9/6/2006.

European Commission, 2008. *Causes and circumstances of accidents at work in the EU*. Belgium: European Commission.

European Commission, 2014. *Mechanical engineering*. [online] European Commission. Available at: <http://ec.europa.eu/cgi-bin/etal.pl> [5 September 2014].

European Commission, n.d. *National provisions communicated by the Member States concerning: directive on machinery*. [online] European Commission. Available at: <http://ec.europa.eu/enterprise/sectors/mechanical/index_en.htm> [27 August 2014].

European Commission, n.d. *Titles and references of harmonised standards.* [online] European Commission. Available at: <http://ec.europa.eu/enterprise/policies/european–standards/harmonised–standards/machinery/index_en.htm> [Last accessed 27 August 2012].

European Committee for Electrotechnical Standardization, 2000. *IEC 60204: 2000 Safety of machinery – electrical equipment of machines* (various parts). Brussels: CENELEC.

European Committee for Standardization, 1997. *EN 1050: 1997 Safety of machinery – principles for risk assessment.* Brussels: CEN.

European Committee for Standardization, 2001a. *EN ISO 14122–3: 2001 Safety of machinery – permanent means of access to machinery – part 3: stairs, stepladders and guard–rails.* Brussels: CEN.

European Committee for Standardization, 2001b. *EN 62079: 2001 Preparation of instructions – structuring, content and presentation.* Brussels: CEN.

European Committee for Standardization, 2003a. *EN ISO 12100–1: 2003 Safety of machinery. Basic concepts, general principles for design. Basic terminology, methodology.* Brussels: CEN.

European Committee for Standardization, 2003b. *EN ISO 12100–2: 2003 Safety of machinery. Basic concepts, general principles for design. Technical principles.* Brussels: CEN.

European Committee for Standardization, 2010. *EN ISO 12100: 2010 Safety of machinery. General principles for design. Risk assessment and risk reduction.* Brussels: CEN.

European Standards Organizations, 2009. *China–EU–EFTA standardization information platform memorandum of understanding (MOU).* Brussels: European Standards Organizations.

Fadier, E. and de la Garza, C., 2006. Safety design: towards a new philosophy. *Safety Science*, 44(1), pp. 55–73.

Fadier, E. and de la Garza, C., 2007. Towards a proactive safety approach in the design process: the case of printing machinery. *Safety Science*, 45(1–2), pp. 99–229.

Fadier, E., de la Garza, C. and Didelot, A., 2003. Safe design and human activity: construction of a theoretical framework from an analysis of a printing sector. *Safety Science*, 41(9), pp. 759–89.

Fairman, R. and Yapp, C., 2005a. Enforced self–regulation, prescription and conceptions of compliance within small business: the impact of enforcement. *Law and Policy*, 27 (Fall), pp. 491–519.

Fairman, R. and Yapp, C., 2005b. HSE RR 366: 2005. *Making an impact on SME compliance behaviour: an evaluation of the effective interventions upon compliance with health and safety legislation in small and medium sized enterprises.* Norwich: HMSO.

Fan, J., Foley, M., Rauser, E. and Siverstein, B., 2006. *SHARP technical report 70–03–2006: The effect of DOSH enforcement inspections and consultation visits on the compensable claims rates in Washington state.* Washington: Washington State Department of Labor and Industries.

Fan, J., Foley, M. and Siverstein, B., 2003. *SHARP technical report 70–3–2003: Impact of WISHA activities on compensable claims rates in Washington State.* Washington: Washington State Department of Labor and Industries.

Federation of European Materials Handling, 2010. *Policy Paper 2009–2014: 2010 Responding to challenges facing the European materials handling industry.* Brussels: FEM.

FEM see Federation of European Materials Handling.

Feyen, R., Liu, Y., Chaffin, D., Jimmerson, G. and Joseph, B., 2000. Computer–aided ergonomics: a case study of incorporating ergonomics analyses into workplace design. *Applied Ergonomics*, 31(3), pp. 291–300.

Fitzharris, M., Yu, J., Hammond, N., Taylor, C., Wu, Y., Finfer, S. and Myburgh, J., 2011. Injury in China: a systematic review of injury surveillance studies conducted in Chinese hospital emergency departments. *BMC Emergency Medicine*, 11(18), pp. 1–16.

Flick, U., 2006. *An introduction to qualitative research.* 3rd ed. London: Sage Publications.

Fraser, I. ed., 2010. Guide to application of the Machinery Directive 2006/42/EC. [online] European Commission. Available at: <http://ec.europa.eu/enterprise/sectors/mechanical/files/machinery/guide_application_directive_2006–42–ec–2nd_edit_6–2010_en.pdf> [Last accessed 27 August 2014].

Frick, K., 2005. European Union's legal standard on risk assessment. In: W. Karwowski ed. 2005. *Handbook of standards and guidelines on ergonomics and human factors.* New Jersey: Lawrence Erlbaum. pp. 471–91.

Frick, K. and Wren, J., 2000. Reviewing occupational health and safety management – multiple roots, diverse perspectives and ambiguous outcomes. In: K. Frick, P.L. Jensen, M. Quinlan and T. Wilthagen eds. 2000. *Systematic occupational health and safety management, perspectives on an international development.* Amsterdam: Pergamon. pp. 17–42.

Gadd, S., Keeley, D. and Balmforth, H., 2003. *HSE RR 151:2003 Good Practice and Pitfalls in Risk Assessment.* Norwich: HMSO.

Gallagher, C., 1997. *Health and safety management systems: an analysis of systems types and effectiveness.* Sydney: National Occupational Health and Safety Commission.

Gardner, D., Cross, J., Fonteyn, D., Carlopio, J. and Shikdar, A., 1999. Mechanical equipment injuries in small manufacturing businesses. *Safety Science*, 33(1–2), pp. 1–12.

Garrigou, A., Daniellou, F., Carballeda, G. and Ruaud, S., 1995. Activity analysis in participatory design and analysis of participatory design activity. *International Journal of Industrial Ergonomics*, 15(5), pp. 311–27.

Genn, H., 1993. Business responses to the regulation of health and safety in England. *Law and Policy*, 15(3), pp. 219–33.

Gergen, K., 1994. *Realities and relationships: soundings in social construction.* Cambridge: Harvard University Press.

Gherardi, S., 2008. Situated knowledge and situated action: what for practice–based studies promise? In: D. Barry and H. Hansen eds. 2008. *The Sage handbook of new approaches in management and organization.* London: Sage. pp. 516–27.

Gigerenzer, G. and Selten, 2001. Rethinking rationality. In: G. Gigerenzer and R. Selten eds. *Bounded rationality, the adaptive toolbox.* Cambridge: MIT Press. pp. 1–12.

Gilad, I. and Reuven, K., 1997. Architecture of an expert system for ergonomics analysis and design. *International Journal of Industrial Ergonomics*, 23(3), pp. 205–21.

Gillham, B., 2000. *The research interview.* London: Continuum.

Glaser, B. and Strauss, A., 1967. *The discovery of grounded theory: strategies for qualitative research.* Chicago: Aldine.

Glendon, I., Clarke, S. and McKenna, E., 2006. *Human safety and risk management.* 2nd ed. Boca Raton: Taylor and Francis.

Gray, W. and Mendeloff, J., 2002. *The declining effects of OSHA inspections on manufacturing injuries: 1979 to 1998.* Massachusetts: National Bureau of Economic Research.

Gray, W. and Scholz, J., 1990. OSHA enforcement and workers' injuries: a behavioural approach to risk assessment. *Journal of Risk Assessment*, 3(3), pp. 283–305.

Gray, W. and Scholz, J. 1991. Analysing the equity and efficiency of OSHA enforcement. *Law and Policy*, 3(3), pp. 185–214.

Gray, W. and Scholz, J., 1993. Does regulatory enforcement work? A panel analysis of OSHA enforcement. *Law Society Review*, 27(1), pp. 177–213.

Green, W. and Jordan, P., 1999. *Human factors in product design: current practices and future trends.* London: Taylor Francis.

Green, W., Kanis, H. and Vermeeren, A., 1997. Tuning the design of everyday products to cognitive and physical activities of users. In S. Robertson ed. 1997. Contemporary ergonomics. London: Taylor and Francis. pp. 175–180.

Gunitalaka, A., 2005. Robots and safety. *OHS Alert*, 9, p. 2.

Gunningham, N., 1984. *Safeguarding the worker. Job hazards and the role of the law.* Sydney: The Law Book Company.

Gunningham, N., 2010. Enforcement and compliance strategies. In: R. Baldwin, M. Cave and M. Lodge M eds. 2010. *The Oxford handbook of regulation.* Oxford: Oxford University Press. pp. 118–45.

Gunningham, N. and Johnstone, R., 1999. *Regulating workplace safety. Systems and sanctions.* Oxford: Oxford University Press.

Gunningham, N., Kagan, R. and Thornton, D., 2003. *Shades of green; business regulation and environment.* Stanford: Stanford University Press.

Gunningham, N. and Sinclair, D., 2002. *Leaders and laggards. Next-generation environmental regulation.* Sheffield: Greenleaf Publishing Ltd.

Gunningham, N., Thornton, D. and Kagan, R., 2005. Motivating management: corporate compliance in environmental protection. *Law and Policy*, 27(2), pp. 289–316.

Haddon, W., 1973. Energy damage and the ten countermeasure strategies. *Journal of Trauma*, 13(4), pp. 321–31.

Haddon, W., 1974. Strategy in preventive medicine: passive versus active approaches to reducing human wastage. *Journal of Trauma*, 14(4), pp. 353–4.

Haddon, W., 1980. The basic strategies for reducing damage from hazards of all kinds. *Hazard Prevention*, 16(11), pp. 8–12.

Haines, F., 1997. *Corporate regulation – beyond 'punish or persuade'*. Oxford: Clarendon Press.

Haines, F., 2011. Paradox of regulation: what regulation can achieve and what it cannot. Cheltenham: Edward Elgar.

Haines, F. and Gurney, D., 2003. The shadows of the law: contemporary approaches to regulation and the problem of regulatory conflict. *Law and Policy*, 25(4), pp. 353–80.

Haines, H. and Wilson, J., 1998. *HSE RR 174: 1998 Development of a framework for participatory ergonomics*. Norwich: HMSO.

Hale, A., 2003. Safety management in production. *Human Factors and Ergonomics in Manufacturing*, 13(3), pp. 185–201.

Hale, A. and Hovden, J., 1998. Management and culture: the third age of safety. A review of approaches to organisational aspects of safety, health and environment. In: A.–M. Feyer and A. Williamson eds. 1998. *Occupational injury. Risk, prevention and intervention*. London: Taylor and Francis. pp. 129–65.

Hale, A. and Swuste, P., 1997. Avoiding square wheels: international experience in sharing solutions. *Safety Science*, 25(1–3), pp. 3–14.

Hale, A., Kirwan, B. and Kjellen, U., 2007. Safe by design: where are we now? *Safety Science*, 45, pp. 305–27.

Hancher, L. and Moran, M., 1989. Organising regulatory space. In: L. Hancher and M. Moran eds. 1989. *Capitalism, culture and economic regulation*. Oxford: Oxford University Press. pp. 271–300.

Hanson, M., Tesh, K., Groat, S., Donnan, P., Ritchie, P., Lancaster, R., 1998. *HSE RR 177: 1998 Evaluation of the six–pack regulations*. Norwich: HMSO.

Harms–Ringdahl, L., 2001. *Safety analysis. principles and practice in occupational safety*. 2nd ed. London: Taylor and Francis.

Harris, J. and Current, R., 2012. Machine safety: new and updated consensus standards. *Professional Safety*, 57(5), pp. 50–7.

Hasan, R., Bernard, A., Ciccotelli, J. and Martin, P., 2003. Integrating safety into the design process: elements and concepts relative to the working situation. *Safety Science*, 41(2–3), pp. 155–79.

Hasle, P., Bager, B. and Granerud, L., 2010. Small enterprises – accountants as occupational health and safety intermediaries. *Safety Science*, 48(3), pp. 404–9.

Hasle, P., Kines, P. and Andersen, L. P., 2009. Small enterprise owners' accident causation attribution and prevention. *Safety Science*, 47(1), pp. 9–19.

Hawkins, K., 2002. *Law as last resort. Prosecution decision making in a regulatory agency*. Oxford: Oxford University Press.

Heads of Workplace Safety Authorities, 2007. *Post implementation report. Agricultural plant designers, manufacturers, suppliers, importers program*. Sydney: Heads of Workplace Safety Authorities.

Healey, N. and Greaves, D., 2007. *A review of consistency of references to risk management frameworks in HSE guidance*. Norwich: HMSO.

Health and Safety Executive, 1997. *Successful health and safety management*. Norwich: HMSO.

Health and Safety Executive, 1998. *INDG 270 04/98 C200: 1998 Supplying new machinery*. Norwich: HMSO.

Health and Safety Executive, 2001a. *Principles and guidelines to assist HSE in its judgements that duty–holders have reduced risk as low as reasonably practicable*. [online] Health and Safety Executive. Available at: <http://www.hse.gov.uk/risk/theory/alarp1.htm> [Last accessed 27 August 2014].

Health and Safety Executive, 2001b. *Reducing risks, protecting people. HSE's decision making process*. Sudbury: HSE Books.

Health and Safety Executive, 2002. *INDG268: 2002 Working together: guidance on health and safety for contractors and suppliers*. Sudbury: HSE Books.

Health and Safety Executive, 2013a. *New machinery*. [online] Health and Safety Executive. Available at: <http://www.hse.gov.uk/work–equipment–machinery/new–machinery.htm> [Last accessed 27 August 2014].

Health and Safety Executive, 2013b. *HSE's role as a market surveillance authority*. [online] Available at: <http://www.hse.gov.uk/work–equipment–machinery/hse–role–market–surveillance–authority.htm> [Last accessed 27 August 2014].

Hopkins, A., 1995. *Making safety work. getting management commitment to occupational health and safety*. Sydney: Allen and Unwin.

Hopkins, A. and Hogan, L., 1998. Influencing small business to attend to occupational health and safety. *Journal of Occupational Health and Safety – Australia and New Zealand*, 14(3), pp. 237–44.

Horlick–Jones T., 2007. On 'risk work': professional discourse, accountability and everyday action. *Health, Risk and Society*, 7(3), pp. 293–307.

Howard, J., 2008. Prevention through design – introduction. *Journal of Safety Research*, 39(2), p. 113.

HSE see Health and Safety Executive.

Hunter, T., 1999. Integrating concepts into the design process. In: W. Christensen and F. Manuele eds. 1999. *Safety through design*. USA: National Safety Council Press. pp. 73–87.

Hutter, B., 1997. *Compliance: regulation and environment*. Oxford: Oxford University Press.

Hutter, B., 2001. *Regulation and risk. Occupational health and safety on the railways*. Oxford: Oxford University Press.

Hutter, B., 2005. Risk management and governance. In: P. Eliadis, M. Hill and M. Howlett eds. 2005. *Designing government: from instruments to governance.* Montreal: McGill–Queens' University Press. pp. 303–21.

Hutter, B., 2006. *CARR Discussion Paper 37: 2006 The role of non–state actors in regulation.* London: The London School of Economics and Political Science.

Hutter, B, 2011. *Managing Food Safety and Hygiene. Governance and Regulation as Risk Management,* Cheltenham, UK and Northampton, MA: Edward Elgar.

Hutter, B. and Jones, C., 2007. From government to governance: external influences on business risk management. *Regulation and Governance,* 1(1), pp. 27–45.

HWSA see Heads of Workplace Safety Authorities.

ILO see International Labour Organization.

IMS Research, 2009. *The European machinery production handbook – 2009.* UK: IMS Research.

Industry Commission, 1995. *Volume 1: 1995 Work, health and safety. Inquiry into occupational health and safety.* Canberra: AGPS.

Industry's Support Platform, 2013a. Market Surveillance. [online] Available at: <http://machinery–surveillance.eu> [Last accessed 27 August 2014].

Industry's Support Platform, 2013b. The 10 key actions for an effective market surveillance. [online] Available at: <http://ec.europa.eu/enterprise/sectors/mechanical/files/machinery/masu–conf/manifesto_en.pdf> [Last accessed 27 August 2014].

International Association of Classification Societies, 2006. *Classification societies – what, why and how?* [online] Available at: <http://www.iacs.org.uk/document/public/explained/Class_WhatWhy&How.PDF > [Last accessed 27 August 2014].

International Labour Organization, 2001. *Guidelines on occupational safety and health management systems.* Geneva: International Labour Organization.

International Organization for Standardization, 2002. *ISO 10535: 2002 Hoists for the transfer of disabled persons. Requirements and test methods.* Geneva: International Organization for Standardization.

International Organization for Standardization, 2009. *ISO 31000: 2009 Risk Management – Principles and Guidelines.* Geneva: International Organization for Standardization.

International Organization for Standardization, 2013. *The international language of ISO graphical symbols.* Geneva: International Organization for Standardization.

ISO see International Organization for Standardization.

James, P., Vickers, I., Smallbone, D. and Baldock, R., 2004. The use of external sources of health and safety information and advice: the case of small firms. *Policy and Practice in Health and Safety,* 2(1), pp. 91–103.

Jamieson, S., Reeve, B., Schofield, T. and McCallum, R., 2010. OHS prosecutions: do they deter other companies from offending? *Journal of Occupational Health, Safety and Environment,* 26(3), pp. 213–31.

Janicik, T., 1999. Designing for maintainability, reliability, and safety. In: W. Christensen and F. Manuele eds. 1999. *Safety through design*. USA: National Safety Council Press. pp. 109–18.

Jensen, P.L., 2001. Risk assessment: a regulatory strategy for stimulating work environment activities? *Human Factors and Ergonomics in Manufacturing*, 11(2), pp. 101–16.

Jensen, P.L., 2002a. Human factors and ergonomics in planning of production. *International Journal of Industrial Ergonomics*, 29(3), pp. 121–31.

Jensen, P.L., 2002b. Assessing assessment: the Danish experience of worker participation in risk assessment. *Economic and Industrial* Democracy, 23(2), pp. 201–27.

Johnstone, R., 1997. *Occupational health and safety law and policy. Text and materials*. 1st ed. Sydney: LBC Information Services.

Johnstone, R., 2003. *Occupational health and safety, courts and crime. The legal construction of occupational health and safety offences in Victoria*. Sydney: The Federation Press.

Johnstone, R., 2004a. *Occupational health and safety law and policy. Text and materials*. 2nd ed. Sydney: Lawbook Co.

Johnstone, R., 2004b. Rethinking OHS enforcement. In: L. Bluff, N. Gunningham and R. Johnstone eds. 2004. *OHS regulation for a changing world of work*. Sydney: The Federation Press. pp. 146–78.

Johnstone, R., Bluff, E. and Clayton, A., 2012. *Work health and safety law and policy*. Sydney: Thomson Reuters.

Johnstone, R. and Jones, N., 2006. Constitutive regulation of the firm: OHS, dismissal, discrimination and sexual harassment. In: C. Arup, P. Gahan, J. Howe, R. Johnstone, R, Mitchell and A. O'Donnell eds. 2006. *Labour law and labour market regulation: essays on the construction, constitution and regulation of labour markets and workplace relationships*. Sydney: The Federation Press. pp. 483–502.

Johnstone, R. and King, M., 2008. A responsive sanction to promote systematic compliance? Enforceable undertakings in occupational health and safety regulation. *Australian Journal of Labour Law*, 21(3), pp. 280–315.

Kagan, R., Gunningham, N. and Thornton, D., 2011. Fear, duty, and regulatory compliance: lessons from three research projects. In: C. Parker and V.L. Nielsen eds. 2011. *Explaining compliance. Business responses to regulation*. Cheltenham: Edward Elgar. pp. 37–58.

Kagan, R. and Scholz, J., 1984. The 'criminology of the corporation' and regulatory enforcement strategies. In: K. Hawkins and L. Thomas eds. 1984. *Enforcing regulation*. Boston: Kluwer. pp. 67–95.

KAN see Kommission Arbeitsschutz und Normung.

Kanis, H., 1998. Usage centred research for everyday product design. *Applied Ergonomics*, 29(1), pp. 75–82.

Karageorgiou, A., Jensen, P.L., Walters, D. and Wilthagen, T., 2000. Risk assessment in four member states of the European Union. In: K. Frick,

P.L. Jensen, M. Quinlan and T. Wilthagen eds. 2000. *Systematic occupational health and safety management. Perspectives on an international development.* Amsterdam: Pergamon. pp. 251–84.

Karwowski, W. and Marras, W., 1999. *The occupational ergonomics handbook.* Florida: CRC Press LLC.

Kerwer, D., 2005. Rules that many use: standard and global regulation. *Governance: An International Journal of Policy, Administration and Institutions*, 18(4), pp. 611–32.

Klein, G., 2001. The fiction of optimization. In: G. Gigerenzer and R. Selten eds. 2001. *Bounded rationality, the adaptive toolbox.* Cambridge: MIT Press. pp. 103–21.

Kletz, T., 1991. *An engineer's view of human error.* Warwickshire: Institution of Chemical Engineers.

Kletz, T., 1998a. Making safety second nature. *Process Safety Progress*, 17(3), pp. 196–9.

Kletz, T., 1998b. *Process plants: a handbook for inherently safer design.* 2ⁿᵈ ed. Philadelphia: Taylor and Francis.

Klinke, A. and Renn, O., 2002a. Precautionary principle and discursive strategies: classifying and managing risks. *Journal of Risk Research*, 4(2), pp. 159–73.

Klinke, A. and Renn, O., 2002b. A new approach to risk evaluation and management: risk–based, precaution–based, and discourse–based strategies. *Risk Analysis*, 22(6), pp. 1071–94.

Ko, K., Mendeloff, J. and Gray, W., 2010. The role of inspection sequence in compliance with the US Occupational Safety and Health Administration's (OSHA) standards: interpretations and implications. *Regulation and Governance*, 4, pp. 48–70.

Kommission Arbeitsschutz und Normung (KAN), 2008. *Safety of agricultural machinery.* Sankt Augustin, Germany: Kommission Arbeitsschutz und Normung.

Konz, S., 2006. Design for ergonomics: facilities planning. In: W. Marras and W. Karwowski eds. 2006. *Interventions, controls, and applications in occupational ergonomics.* 2nd ed. Boca Raton: Taylor and Francis. Ch.11.

Kouabenen, D., Gilbert, D., Médina, M. and Bouzon, F., 2001. Hierarchical position, gender, accident severity, and causal attribution. *Journal of Applied Social Psychology*, 31(3), pp. 553–75.

Kuorinka, I., 1997. Tools and means of implementing participatory ergonomics. *International Journal of Industrial Ergonomics*, 19(4), pp. 267–70.

Lacore, J.–P., 2002. From EN 292 to EN ISO 12100: developments in safe machinery design principles from 1985 to 2002. In: T. Koukoulaki and S. Boy eds. 2002. *Globalising technical standards. Impact and challenges for Occupational health and safety.* Brussels: European Trade Union Technical Bureau for Health and Safety. pp. 39–56.

Lambert, J. and Associates, 2003. *Forklift stability and other technical safety issues.* Melbourne: Monash University Accident Research Centre.

Lamm, F. and Walters, D., 2004.Regulating occupational health and safety in small business. In: L. Bluff, N. Gunningham and R. Johnstone eds. 2004. *OHS regulation for a changing world of work*. Sydney: The Federation Press. pp. 94–119.

Lave, J. and Wenger, E., 1990. *Situated learning: legitimate peripheral participation*. Cambridge: Cambridge University Press.

Licht, A., 2008. *Expanded rationality: from the preferred to the desirable, with some implications for law*. Israel: Radzyner School of Law.

Luntz, H. and Hambly, D., 2002. *Torts. Cases and commentary*. 5th ed. Sydney: LexisNexis Butterworths.

Main, B., 1999. Applying concepts to product liability prevention. In: W. Christensen and F. Manuele eds. 1999. *Safety through design*. USA: National Safety Council Press. pp. 139–51.

Manuele, F., 1999a. Why safety through design: what's in it for you? In: W. Christensen and F. Manuele eds. 1999. *Safety through design*. USA: National Safety Council Press. pp. 3–8.

Manuele, F., 1999b. Concepts, principles and methods for safety through design. In: W. Christensen and F. Manuele eds. 1999. *Safety through design* USA: National Safety Council Press. pp. 9–21.

Manuele, F., 2008. Prevention through Design (PtD): history and future. *Journal of Safety Research*, 39(2), pp. 127–30.

Marras, W. and Karwowski, W. eds, 2006. *Interventions, controls, and applications in occupational ergonomics*. 2nd ed. Boca Raton: Taylor and Francis.

Marshall, C. and Rossman, G., 2006. *Designing qualitative research*. 4th ed. Thousand Oaks: Sage Publications.

Mason, J., 1996. *Qualitative researching*. London: Sage Publications.

Maxwell, C., 2004. *Occupational Health and Safety Act review*. Melbourne: State of Victoria.

May, P., 2004. Compliance motivations: affirmative and negative bases. *Law and Society Review*, 38(1), pp. 41–68.

May, P. and Wood, R., 2003. At the regulatory frontlines: inspectors' enforcement styles and regulatory compliance. *Journal of Public Administration Research and Theory*, 13(2), pp.117–139.

Mayhew, C., 1997a. *Barriers to implementation of known OHS solutions in small business*, Canberra: AGPS.

Mayhew, C., 1997b. Small business occupational health and safety provision. *Journal of Occupational Health and Safety – Australia and New Zealand*, 13(4), pp. 361–73).

Mayhew, C., Young, C., Ferris, R. and Harnett, C., 1997. *An evaluation of the impact of targeted interventions on the ohs behaviours of small business building industry owners/managers/contractors*. Sydney: National Occupational Health and Safety Commission.

Melrose, A., Graham, M., Graveling, R., George, J., Cowie, H., Hutchison, P. and Mulholland, R., 2006. *HSE RR 486: 2006 Assessing the effectiveness of the manual handling assessment chart (MAC) and supporting website.* Norwich: HMSO.

Mendeloff, J. and Gray, W., 2005. Inside the black box: how do OSHA inspections lead to reductions in workplace injuries? *Law and Policy*, 27(2), pp. 219–37.

Miles, M. and Huberman, A. M., 1994. *An expanded sourcebook. qualitative data analysis.* 2nd ed. Thousand Oaks: Sage Publications.

Mills, S., 2000. The importance of task analysis in usability context analysis – designing for fitness for purpose. *Behaviour and Information Technology*, 19(1), pp. 57–68.

Minichiello, V., Aroni, R., Timewell, E. and Alexander, L., 1995. *In–depth interviewing. principles, techniques, analysis.* 2nd ed. Sydney: Longman.

Morgan, B. and Yeung, K., 2007. *An introduction to law and regulation.* Cambridge: Cambridge University Press.

Morris, W., Wilson, J. and Koukoulaki, T., 2004. *Developing a participatory approach to the design of work equipment. assimilating lessons from workers' experience.* Brussels: European Trade Union Technical Bureau for Health and Safety.

Morse, J. and Richards, L., 2002. *Read me first for a user's guide to qualitative methods.* Thousand Oaks: Sage Publications.

Murphy, K., 2005. Regulating more effectively: the relationship between procedural justice, legitimacy and tax non–compliance. *Journal of Law and Society*, 32(4), pp. 562–89.

Nachreiner, F., Nickel, P. and Meyer, I., 2006. Human factors in process control systems: the design of human–machine interfaces. *Safety Science*, 44(1), pp. 5–26.

National Institute for Occupational Safety and Health, 2006. *Prevention through design.* [online] Available at: <http://www.cdc.gov/niosh/topics/PtD> [Last accessed 27 August 2014].

National Institute for Occupational Safety and Health, 2013. *DHHS (NIOSH) Publication 2013–134: PtD Mechanical – Electrical Systems Instructors' Manual.* Ohio: Department of Health and Human Services.

National Occupational Health and Safety Commission, 1994. *NOHSC 1010: 1994 National standard for plant.* Canberra: AGPS.

National Occupational Health and Safety Commission, 1998. *NOHSC 7025: 1998 National guidelines for integrating occupational health and safety competencies into national industry competency standards.* Canberra: National Occupational Health and Safety Commission.

National Occupational Health and Safety Commission, 2000. *Work–related fatalities associated with design issues involving machinery and fixed plant in Australia, 1989 to 1992.* Sydney: National Occupational Health and Safety Commission.

National Occupational Health and Safety Commission, 2002. *National OHS strategy 2002–2012.* Canberra: National Occupational Health and Safety Commission.

National Occupational Health and Safety Commission, 2005. *Work–related fatalities study team – reports and guidance relating to particular plant.* Canberra: National Occupational Health and Safety Commission.

Neathey, F., Sinclair, A., Rick, J., Ballarad, J., Hunt, W. and Denvir, A., 2006. *HSE RR 476: 2006 Evaluation of the five steps to risk assessment.* Norwich: HMSO.

Neboit, M., 2003. A support to prevention integration since design phase: the concepts of 'limit conditions' and 'limit activities' tolerated by use. *Safety Science*, 41(2–3), pp. 95–109.

Neuman, W. L., 1997. *Social research methods. Qualitative and quantitative approaches.* 3rd ed. Boston: Allyn and Bacon.

NIOSH see National Institute for Occupational Safety and Health (United States)

NOHSC see National Occupational Health and Safety Commission.

Nytrö, K., Saksvik, P.O. and Torvatn, H., 1998. Organisational prerequisites for the implementation of systematic health, environment and safety work in enterprises. *Safety Science*, 30(3), pp. 297–307.

OSHA see Occupational Safety and Health Administration.

Occupational Safety and Health Administration, 2014. *Machine guarding.* [online] Available at: <https://www.osha.gov/SLTC/machineguarding/> [Last accessed 27 August 2014].

Ottosson, S., 2002. Virtual reality in the product development process. *Journal of Engineering Design*, 13(2), pp. 159–72.

Palincsar, A. S., 1998. Social constructivist perspectives on teaching and learning. *Annual Review of Psychology*, 49(1), pp. 345–75.

Paquet, V. and Lin, L., 2003. An integrated methodology for manufacturing systems design using manual and computer simulation. *Human Factors and Ergonomics in Manufacturing*, 13(1), pp. 19–40.

Parker, C., 2002. *The open corporation. Effective self–regulation and democracy.* Cambridge: Cambridge University Press.

Parker, C. and Gilad, S., 2011. Internal corporate compliance management systems: structure, culture and agency. In: C. Parker and V. L. Nielsen eds. 2011. *Explaining compliance. Business responses to regulation.* Cheltenham: Edward Elgar. pp. 170–95.

Parker, C. and Nielsen, V. L., 2009. The challenge of empirical research on business compliance in regulatory capitalism. *Annual Review of Law and Social Sciences*, 5, pp. 45–70.

Parker, C. and Nielsen, V. L., 2011. *Explaining compliance. Business responses to regulation.* Cheltenham: Edward Elgar.

Pappas, M., Karabatsou, V., Mavrikios, D. and Chryssolouris, G., 2007. Ergonomic evaluation of virtual assembly tasks. In: P. Cunha and P. Maropoulos eds. 2007. *Digital enterprise technology. Perspectives and future challenges.* New York: Springer. pp. 511–8.

Pearse, W., 2001. Club zero: implementing OHS management systems in small to medium fabricated metal product companies. In: W. Pearse, C. Gallagher and L. Bluff eds. 2001. *Occupational health and safety management systems.* Melbourne: Crown Content. pp. 83–100.

Polet, P., Vanderhaegen, F. and Amalberti, R., 2003. Modelling border-line tolerated conditions of use (BTCU) and associated risks. *Safety Science*, 41(2–3), pp. 111–36.

Productivity Commission, 2006. *Standard setting and laboratory accreditation.* Canberra: Commonwealth of Australia.

Raafat, H., 1989. Risk assessment and machinery safety. *Journal of Occupational Accidents*, 11(1), pp. 37–50.

Raafat, H. and Sadhra, S., 1999. Risk characterisation. In: S. Sadhra and K. Rampal eds. 1999. *Occupational health. Risk assessment and management.* London: Blackwell Science. Ch. 10.

Raafat, H. and Simpson, P., 2001. Integrating safety during the machine design stage. *Safety Through Design*, June, pp. 1–7.

Reason, J., 1990. *Human error.* Cambridge: Cambridge University Press.

Reason, J., 1997. *Managing the risks of organisational accidents.* Aldershot: Ashgate.

Reber, A. and Reber, E., 2001. *Dictionary of psychology.* 3rd ed. London: Penguin Books.

Reunanen, M., 1993. *Systematic safety consideration in product design.* Espoo: Safety Engineering Laboratory, Technical Research Centre of Finland.

Richards, L., 2005. *Handling qualitative data. A practical guide.* London: Sage Publications.

Ringelberg, J. and Voskamp, P., 1996. *Integrating ergonomic principles into C–standards for machinery design. TUTB proposals for guidelines.* Brussels: European Trade Union Technical Bureau for Health and Safety.

Robens (Lord), 1972. *Safety and health at work. Report of the committee 1970–1972.* London: HMSO.

SAA see Standards Association of Australia.

Safe Work Australia, 2009. *Notified fatalities statistical report 2008–09.* Canberra: Commonwealth of Australia.

Safe Work Australia, 2011. *Notified fatalities statistical report 2008–09.* Canberra: Commonwealth of Australia.

Safe Work Australia, 2012a. *Australian work health and safety strategy 2012–2022.* Canberra: Safe Work Australia.

Safe Work Australia, 2012b. *Model code of practice – managing the risks of plant in the workplace.* Canberra: Safe Work Australia.

Safe Work Australia, 2013a. *Work–related injuries resulting in hospitalisation, July 2006 to June 2009.* Canberra: Safe Work Australia.

Safe Work Australia, 2013b. *Model work health and safety laws.* [online] Available at: <http://www.safeworkaustralia.gov.au/sites/swa/model–whs–laws/pages/model–whs–laws [Last accessed 27 August 2014].

Safe Work Australia, 2013c. *QuadWatch.* [online] Available at: <http://www.safeworkaustralia.gov.au/sites/swa/whs–information/agriculture/quad–watch/pages/quad–watch> [Last accessed 27 August 2014].

Sagot, J., Gouin, V. and Gomes, S., 2003. Ergonomics in product design: safety factors. *Safety Science* 41(2), pp. 137–54.

Saksvik, P.Ø. and Quinlan, M., 2003. Regulating systematic occupational health and safety management. Comparing the Norwegian and Australian experience. *Industrial Relations*, 58(1), pp. 33–59.

Sanders, M. and McCormick, E., 1993. *Human factors in engineering and design.* 7th ed. New York: McGraw Hill Inc.

Schulte, P., Rinehart, R., Okun, A., Geraci, C. and Heidel, D., 2008. National Prevention through Design (Ptd) initiative. *Journal of Safety Research*, 39(2), pp. 115–21.

Schupp, B., Hale, A., Pasman, H., Lemkovitz, S. and Goossens, L., 2006. Design support with a systematic integration of risk reduction into early chemical process design. *Safety Science*, 44, pp. 37–54.

Schwartz, S. and Bilsky, W., 1987. Towards a universal psychological structure of human values. *Journal of Personality and Social Psychology*, 53, pp. 550–62.

Scott, C., 2010. Standard–setting in regulatory regimes. In: R. Baldwin, M. Cave and M. Lodge eds. 2010. *The Oxford handbook of regulation.* Oxford: Oxford University Press. pp. 104–19.

Scribner, S. and Beach, K., 1993. An activity theory approach to memory. *Applied Cognitive Psychology,* 7(3), pp. 185–190.

Seim, R. and Broberg, O., 2010. Participatory workspace design: a new approach for ergonomists. *International Journal of Industrial Ergonomics*, 40(1), pp. 25–33.

Selten, R., 2001. What is bounded rationality? In: G. Gigerenzer and R. Selten eds. 2001. *Bounded rationality, the adaptive toolbox.* Cambridge: MIT Press. pp. 13–36.

Shah, S., Silverstein, B. and Foley, M., 2003. *SHARP Technical Report 70–2–200: Workplace health and safety regulations: impact of enforcement and consultation on workers compensation claims rates in Washington state with two years follow–up. Washington:* Washington State Department of Labor and Industries.

Silverman, D., 2001. *Interpreting qualitative data. Methods for analysing talk, text and interaction.* 2nd ed. London: Sage Publications.

Simon, H., 1955. A behavioral model of rational choice. *The Quarterly Journal of Economics*, 69(1), pp. 99–118.

Skinner, W. and Stewart, A., 2006. *Evaluation of the forklift instability and traffic management project #979, July 2005–June 2006.* Melbourne: Victorian WorkCover Authority.

Sparrow, M., 2000. *The regulatory craft. Controlling risks, solving problems, and managing compliance.* Washington: The Brookings Institution.

Standards Australia, 1981. *AS 1250: 1981 The use of steel in structures.* Sydney: Standards Australia.

Standards Australia, 1987. *AS 2939: 1987 Industrial robot systems – safe design and usage.* Sydney: Standards Australia.

Standards Australia, 1992. *AS 1657: 1992 Fixed platforms, walkways, stairways and ladders*. Sydney: Standards Australia.

Standards Australia, 1994a. *AS 1219: 1994 Power presses – safety requirements*. Sydney: Standards Australia.

Standards Australia, 1994b. *AS 1554: 1994 SAA Structural steel welding code*. Sydney: Standards Australia.

Standards Australia, 1994c. *AS 1319: 1994 Safety signs for the occupational Environment*. Sydney: Standards Australia.

Standards Australia, 1996. *AS 4024: 1996 Safeguarding of machinery*. Sydney: Standards Australia.

Standards Australia, 1997. *AS/NZS 4804: 1997 Occupational health and safety management systems – general guidelines on principles, systems and supporting techniques*. Sydney: Standards Australia and Wellington: Standards New Zealand.

Standards Australia, 1998. *AS/NZS 3931: 1998 Risk analysis of technological systems – application guide*. Sydney: Standards Australia and Wellington: Standards New Zealand.

Standards Australia, 1999. *AS/NZS 4360: 1999 Risk management*. Sydney: Standards Australia and Wellington: Standards New Zealand.

Standards Australia, 2000a. *AS 1755: 2000 Conveyors – safety requirements*. Sydney: Standards Australia.

Standards Australia, 2000b. *AS/NZS 1200: 2000 Pressure equipment*. Sydney: Standards Australia and Wellington: Standards New Zealand.

Standards Australia, 2000c. *AS/NZS 2153.1: 2000 Tractors and machinery for agriculture and forestry. Technical means for ensuring safety – general*. Sydney: Standards Australia and Wellington: Standards New Zealand.

Standards Australia, 2000d. *AS 3000: 2000 SAA wiring rules*. Sydney: Standards Australia.

Standards Australia, 2001a. *AS/NZS 2865: 2001 Safe working in a confined space*. Sydney: Standards Australia and Wellington: Standards New Zealand.

Standards Australia, 2001b. AS 4801: 2001 *Occupational health and safety management systems – specification with guidance for use*. Sydney: Standards Australia and Wellington: Standards New Zealand.

Standards Australia, 2002. *AS 1418.1: 2002 Cranes, hoists and winches. Part 1: general requirements*. Sydney: Standards Australia.

Standards Australia, 2004a. *AS/NZS 4360 Risk management*. Sydney: Standards Australia and Wellington: Standards New Zealand.

Standards Australia, 2006a. *AS 4024.1201: 2006 Safety of machinery. General principles – basic terminology and methodology*. Sydney: Standards Australia.

Standards Australia, 2006b. AS 4024.1301: 2006 *Safety of machinery – principles of risk assessment*. Sydney: Standards Australia.

Standards Australia, 2006c. *AS 4024.1601:2006 Safety of machinery – design of controls, interlocks and guarding*. Sydney: Standards Australia.

Standards Australia, 2008a. *Product safety framework. Part 3.5 product safety warning labels and markings*. Sydney: Standards Australia.

Standards Australia, 2008b. *Product safety framework. Part 3.24. informative and instructive material*. Sydney: Standards Australia.

Standards Australia, 2012. *Developing standards*. [online] Available at: <http://www.standards.org.au/DevelopingStandards.aspx> [Last accessed 27 August 2014].

Standards Australia, 2014. *AS 4024 (Parts 1201, 1303): 2014 Safety of machinery*. Sydney: Standards Australia.

Stanton, N. and Young, M., 1999. *A guide to methodology in ergonomics: designing for human use*. London: Taylor and Francis.

Stanton, N., Salmon, P., Walker, G., Baber, C. and Jenkins, D., 2005. *Human factors methods. A practical guide for engineering and design*. Aldershot: Ashgate.

Sundin, S. and Medbo, L., 2003. Computer visualisation and participatory ergonomics as methods in workplace design. *Human Factors and Ergonomics in Manufacturing*, 13(1), pp. 1–17.

Sundström–Frisk C, 1996. Promoting safe behavior. In: *For a Good Working Life*. Stockholm: International Congress on Occupational Health.

Sundström–Frisk, C., 1999. Understanding human behaviour: a necessity in improving safety and health performance. *Journal of Occupational Health and Safety – Australia and New Zealand*, 15(1), pp. 37–45.

Suri, J. and Marsh, M., 2000. Scenario building as an ergonomics method in consumer product design. *Applied Ergonomics*, 31(2), pp. 151–7.

SUT see Swinburne University of Technology.

SWEA see Swedish Working Environment Authority.

Swedish Working Environment Authority, 2002. *Scrutiny of products and documentation – 'market control'*. Stockholm: Work Environment Authority Unit for Machinery and Personal Protective Equipment.

Swinburne University of Technology, 2009. *Engineering Program and Course Outline*. Melbourne: Swinburne University of Technology.

Swuste, P., 1997. Editorial safety in design. *Safety Science*, 26(12), pp. 61–3.

Swuste, P., Goossens, L., Bakker, F. and Schrover, J., 1997. Evaluation of accident scenarios in a Dutch steel works using a hazard and operability study. *Safety Science*, 26(1–2), pp. 63–74.

Swuste, P., Hale, A. and Zimmerman, G., 1997. Sharing workplace solutions by solution data banks. *Safety Science*, 26(1–2), pp. 95–104.

Swuste, P., van Drimmelen, D. and Burdorf, A., 1997. Application of design analysis to solution generation: hand–arm vibration in foundation pile head removal in the construction industry. *Safety Science*, 27(2–3), pp. 85–98.

Thornton, N., Gunningham, N. and Kagan, R., 2005. General deterrence and corporate environmental behaviour. *Law and Policy*, 27(2), pp. 262–88.

Thornton, D., Kagan, R. and Gunningham, N., 2009. When social norms and pressures are not enough: environmental performance in the trucking industry. *Law and Society Review*, 43(2), pp. 405–35.

Toft, Y., Howard, P. and Jorgensen, D., 2003. Changing paradigms for professional engineering practice towards safe design – an Australian perspective. *Safety Science*, 41(2–3), pp. 263–76.

Tractor and Machinery Association, 2001a. *News Release*. Press release, 6 June 2001.

Tractor and Machinery Association, 2001b. *Call to ban sales of farm machinery in New South Wales*. Press release, 27 July 2001.

Tractor and Machinery Association, *Arbor, where to from here*. Press release, 10 October 2001.

Tractor and Machinery Association, *Sometimes it is best not to sell the product*. Press release, 8 March 2002.

Tyler, T., 1997. The psychology of legitimacy: a relational perspective on voluntary deference to authorities. *Personality and Social Psychology Review*, 1(4), pp. 323–45.

Tyler, T., 2001. Trust and law abidingness: a proactive model of social regulation. *Boston University Law Review*, 81(2), pp. 361–406.

Tyler, T., 2006. *Why people obey the law*. Princeton: Princeton University Press.

Vanderkruk, R.,1999. Workplace health and safety officers: a Queensland success story. *Journal of Occupational Health and Safety – Australia and New Zealand*, 16(6), pp. 557–63.

Victorian Trades Hall Council, 2004. *The view from the frontline: a report on the experience of OHS reps*. Melbourne: Victorian Trades Hall Council.

Victorian WorkCover Authority, 1995. *Understanding the plant safety package*. Melbourne: Victorian WorkCover Authority.

Victorian WorkCover Authority, 2000. *Strategy 2000*. Melbourne: Victorian WorkCover Authority.

Victorian WorkCover Authority, 2002a. *WorkSafe Victoria inspectors*. Melbourne: Victorian WorkCover Authority.

Victorian WorkCover Authority, 2002b. *SafetyMAP: auditing health and safety management systems*. 4th ed. Melbourne: Victorian WorkCover Authority.

Victorian WorkCover Authority, 2003a. *2003 annual report, working together.* Melbourne: Victorian WorkCover Authority.

Victorian WorkCover Authority, 2003b. *Field operations manual*. Melbourne: Victorian WorkCover Authority.

Victorian WorkCover Authority, 2003c. *2003–2004 industry levy rates*. Melbourne: Victorian WorkCover Authority.

VTHC see Victorian Trades Hall Council.

Vygotsky, L., 1978. *Mind in society*. London: Harvard University Press.

Walker, D. and Tait, R., 2004. Health and safety management in small enterprises: an effective low cost approach. *Safety Science*, 42(1), pp. 69–83.

Walters, D., 2001. *Health and safety in small enterprises: European strategies for managing improvement*. Brussels: PIE–Peter Lang.

Walters, D., 2002. *Working safely in small enterprises in Europe. Towards a sustainable system for worker participation and representation*. Brussels: European Trade Union Confederation.

Walters, D. and Frick, K., 2000. Worker participation and the management of occupational health and safety: reinforcing or conflicting strategies? In: K. Frick, P.L. Jensen, M. Quinlan and T. Wilthagen eds. 2000. *Systematic occupational health and safety management. Perspectives on an international development.* Amsterdam: Pergamon. pp. 43–65.

Walters, D. and James, P., 2009. *Understanding the role of supply chains in influencing health and safety at work.* Leicester: Institution of Occupational Safety and Health.

Walters, D. and Jensen, P.L., 2000. The discourses and purposes behind the development of the EU Framework Directive. In: K. Frick, P.L. Jensen, M. Quinlan and T. Wilthagen eds. 2000. *Systematic occupational health and safety management. Perspectives on an international development.* Amsterdam: Pergamon. pp. 87–98.

Weegels, M. and Kanis, H., 2000. Risk perception in consumer product use. *Accident Analysis and Prevention,* 32(3), pp. 365–70.

Weil, D., 1996. If OSHA is so bad, why is compliance so good? *RAND Journal of Economics,* 27(3), pp. 618–40.

Weil, D., 2001. Assessing OSHA performance: new evidence from the construction industry *Journal of Policy Analysis and Management,* 20(4), pp. 651–74.

Wettig, J., 2002. New developments in standardisation in the past 15 years – product versus process related standards. *Safety Science,* 40, pp. 51–6.

WHSQ see Workplace Health and Safety Queensland.

Wiseman, J. and Gilbert, F., 2002. *HSE RR 434: 2002 COSHH essentials: survey of firms purchasing this guidance.* Norwich: HMSO.

WorkCover Corporation, 2000. *2000–2001 industry levy rates.* Adelaide: WorkCover Corporation.

WorkCover Corporation, 2001a. *Code for the conduct of exempt employers under the WorkCover scheme.* Adelaide: WorkCover Corporation.

WorkCover Corporation, 2001b. *Safety achiever business system. Performance standards.* Adelaide: WorkCover Corporation.

WorkCover Corporation, 2008. *Safety achiever business system practice guidelines.* Adelaide: WorkCover Corporation.

WorkCover Corporation, 2013. *Code of conduct for self–insured employers under the workcover scheme.* Adelaide: WorkCover Corporation.

WorkCover NSW, 2007. *Grain augers. Industry safety standard.* Sydney: WorkCover NSW.

WorkCover NSW, 2008. *Post drivers. Industry safety standard.* Sydney: WorkCover NSW.

Workplace Health and Safety Queensland, 2006. *Guide to safeguarding common machinery and plant.* Brisbane: Department of Industrial Relations.

Workplace Relations Ministers Council, 2002. *Comparative performance monitoring report.* 4th ed. Canberra: Department of Employment and Workplace Relations.

Workplace Relations Ministers' Council, 2006. *Comparative performance monitoring report.* 8th ed. Canberra: Department of Employment and Workplace Relations.

Workplace Services, 1998a. *Project brief. Safer at the source (plant)*. Adelaide: Department for Administrative and Information Services.

Workplace Services, 1998b. *Project plan – safer at the source*. Adelaide: Department for Administrative and Information Services.

Workplace Services, 1999. *Assistance to suppliers, importers, manufacturers and designers of plant and equipment*. Adelaide: Department for Administrative and Information Services.

Workplace Services, 2000. *Designers, manufacturers, importers and suppliers' OHS responsibilities for plant for use in workplaces*. Adelaide: Department for Administrative and Information Services.

Workplace Services, 2002a. *Machine guarding – major workplace hazards strategy*. Adelaide: Department for Administrative and Information Services.

Workplace Services, 2002b. *Machine guarding, guidance booklet*. Adelaide: Department for Administrative and Information Services.

Workplace Services, 2002c. *Safeguards index*. Adelaide: Department for Administrative and Information Services.

Worksafe Australia, 1996. *Economic impact analysis on the National Standard for Plant*. Canberra: AGPS.

WorkSafe Victoria, 2002. *Plant hazard checklist*. Melbourne: Victorian WorkCover Authority.

WorkSafe Victoria, 2003a. *Forklift safety – reducing the risk*. Melbourne: Victorian WorkCover Authority.

WorkSafe Victoria, 2003b. *Guidance note, forklifts – instability and excessive speed*. Melbourne: Victorian WorkCover Authority.

WorkSafe Victoria, 2008. *SafetyMAP: measuring health and safety management*. Melbourne: WorkSafe Victoria.

Worringham, C., 2004. Incompatible and unsafe: control design and use. In: Safety Institute of Australia, *Proceedings of Visions Conference*. Gold Coast, Australia, 2004. Brisbane: Safety Institute of Australia.

Wright, M., Marsden, S. and Antonelli, A., 2004. *HSE RR 196: 2004 Building an evidence base for the health and safety commission strategy to 2010 and beyond: a literature review of interventions to improve health and safety compliance*. Norwich: HMSO.

Wright, M., Marsden, S., Antonelli, A., Norton, J., Doyle, J., Genna, R. and Bendiq, M., 2005. *HSE RR 334: 2005 An evidence–based evaluation of how best to secure compliance with health and safety law*. Norwich: HMSO.

WRMC see Workplace Relations Ministers' Council.

Yeung, K., 2004. *Securing compliance. A principled approach*. Oxford: Hart Publishing.

Zwetsloot, G., 2000. Developments and debates on OHSM system standardisation and certification. In: K. Frick, P.L. Jensen, M. Quinlan and T. Wilthagen eds. 2000. *Systematic occupational health and safety management. Perspectives on an international development*. Amsterdam: Pergamon. pp. 391–412.

# Index

Note: **Bold** page numbers indicate figures, *italic* numbers indicate tables.